POU/POE Implementation Feasibility Study for Arsenic Treatment

Awwa
Research
Foundation

About the Awwa Research Foundation

The Awwa Research Foundation (AwwaRF) is a member-supported, international, nonprofit organization that sponsors research to enable water utilities, public health agencies, and other professionals to provide safe and affordable drinking water to consumers.

The Foundation's mission is to advance the science of water to improve the quality of life. To achieve this mission, the Foundation sponsors studies on all aspects of drinking water, including supply and resources, treatment, monitoring and analysis, distribution, management, and health effects. Funding for research is provided primarily by subscription payments from approximately 1,000 utilities, consulting firms, and manufacturers in North America and abroad. Additional funding comes from collaborative partnerships with other national and international organizations, allowing for resources to be leveraged, expertise to be shared, and broad-based knowledge to be developed and disseminated. Government funding serves as a third source of research dollars.

From its headquarters in Denver, Colorado, the Foundation's staff directs and supports the efforts of more than 800 volunteers who serve on the board of trustees and various committees. These volunteers represent many facets of the water industry, and contribute their expertise to select and monitor research studies that benefit the entire drinking water community.

The results of research are disseminated through a number of channels, including reports, the Web site, conferences, and periodicals.

For subscribers, the Foundation serves as a cooperative program in which water suppliers unite to pool their resources. By applying Foundation research findings, these water suppliers can save substantial costs and stay on the leading edge of drinking water science and technology. Since its inception, AwwaRF has supplied the water community with more than $300 million in applied research.

More information about the Foundation and how to become a subscriber is available on the Web at **www.awwarf.org.**

POU/POE Implementation Feasibility Study for Arsenic Treatment

Prepared by:
Ramesh Narasimhan, **Sudheera Addepally**, and **Kate Martin**
Narasimhan Consulting Services Inc., Phoenix, AZ 85012

Sunil Kommineni
Malcolm Pirnie Inc., Phoenix, AZ 85008

and

Jerry Lowry
Lowry Environmental Engineering, Blue Hill, ME 04614

Sponsored by:
Arsenic Research Partnership

Jointly funded by:
Awwa Research Foundation
6666 West Quincy Avenue, Denver, CO 80235-3098

and

United States Environmental Protection Agency
Washington, D.C. 20460

Published by:

Awwa Research Foundation
Advancing the Science of Water®

DISCLAIMER

CONTENTS

TABLES

FIGURES

FOREWORD

The Awwa Research Foundation (AwwaRF) is a nonprofit corporation that is dedicated to the implementation of a research effort to help utilities respond to regulatory requirements and traditional high-priority concerns of the industry. The research agenda is developed through a process of consultation with subscribers and drinking water professionals. Under the umbrella of a Strategic Research Plan, the Research Advisory Council prioritizes the suggested projects based upon current and future needs, applicability, and past work; the recommendations are forwarded to the Board of Trustees for final selection. The foundation also sponsors research projects through an unsolicited proposal process; the Collaborative Research, Research Applications, and Tailored Collaboration programs; and various joint research efforts with organizations such as the U.S. Environmental Protection Agency, the U.S. Bureau of Reclamation, and the Association of California Water Agencies.

This publication is a result of one of these sponsored studies, and it is hoped that its findings will be applied in communities throughout the world. The following report serves not only as a means of communicating the results of the water industry's centralized research program but also as a tool to enlist the further support of the nonmember utilities and individuals.

Projects are managed closely from their inception to the final report by the foundation's staff and large cadre of volunteers who willingly contribute their time and expertise. The foundation serves a planning and management function and awards contracts to other institutions such as water utilities, universities, and engineering firms. The funding for this research effort comes primarily from the Subscription Program, through which water utilities subscribe to the research program and make an annual payment proportionate to the volume of water they deliver and consultants and manufacturers subscribe based on their annual billings. The program offers a cost-effective and fair method for funding research in the public interest.

A broad spectrum of water supply issues is addressed by the foundation's research agenda: resources, treatment and operations, distribution and storage, water quality and analysis, toxicology, economics, and management. The ultimate purpose of the coordinated effort is to assist water suppliers to provide the highest possible quality of water economically and reliably. The true benefits are realized when the results are implemented at the utility level. The foundation's trustees are pleased to offer this publication as a contribution toward that end.

AwwaRF Project 2730, *POU/POE Implementation Feasibility Study for Arsenic Treatment*, included a review of previously conducted POU and POE treatment studies and federal and state regulations. Information from a survey of state regulators provided further insight into the current status of state rules and guidance in place for POU and POE treatment. A field screening evaluation was conducted at six different participating utility sites to assess the performance of several POU and POE technologies. The performance of POU units was further evaluated in testing at homes in communities served by three different water systems. The findings from the review of previous studies, regulatory review, field screening evaluation, and in-home POU testing formed the basis for developing guidelines for implementing a POU or POE treatment program for arsenic compliance.

Walter J. Bishop
Chair, Board of Trustees
Awwa Research Foundation

Robert C. Renner
Executive Director
Awwa Research Foundation

ACKNOWLEDGMENTS

The authors would like to acknowledge the generous financial support of AwwaRF, ACWA, and USEPA, without which this study would not have been possible. The development of this work has been overseen from the beginning by an enthusiastic Project Manager and experienced Project Advisory Committee (PAC) whose guidance, expertise and suggestions were of valuable assistance. The project team would like to thank the following Project Managers and the PAC members for their kind support, time, and effort:

Project Managers:	Ms. Misha Hasan and Ms. India Williams, AwwaRF, Denver, Colo.
PAC Members:	Mr. Joseph Harrison, Water Quality Association, Lisle, Ill.
	Mr. Jeanne Bailey, Fairfax County Water Authority
	Mr. Jeff Kempic, OGWDW, USEPA
	Mr. Mark Benjamin, University of Washington, Seattle, Wash.

Several utilities across the U.S. participated in this project by coordinating field testing activities, performing analytical testing, supplying their utility information for report writing, and providing valuable comments from time to time. The authors would like to acknowledge the support of the following water utility participants:

Frank Walker, Water Quality Laboratory, Tucson Water Department, Ariz.
Tom Jefferson, Fawad Mohammad, Mike Dew, Jason Saline, Charlie Carroll, Tucson Water Department, Ariz.
Marlon Cook, Stagecoach General Improvement District, Stagecoach, Nev.
Troy Day, Arizona American Water Company, Ariz.
Curtis Horton, Carson City Utilities Department, Nev.
Chris Hill, Metro Water District, Ariz.
Jeff Whitney, Sandy Stream Park, Unity, Maine

Several manufacturers and vendors of POU and POE treatment devices kindly donated equipment, services, and technical advice to the project. The project team would like to acknowledge the support of the following individuals and companies:

Sam Mason, Keith Brown, Kinetico
Wei C. Chen, Apyron
Peter Erikson, Osmonics
Bob Johnson Associates, Inc., Houston, Texas

Finally, the authors would like to acknowledge the support and services provided by other members of the research team:

Bruce Thompson, University of New Mexico, N.M.
Aaron Dotson, Malcolm Pirnie, Inc.
Herb Durbin, Narasimhan Consulting Services, Inc.
Greg Harrison, Narasimhan Consulting Services, Inc.
Dennis Clifford, University of Houston, Texas

EXECUTIVE SUMMARY

The recently promulgated arsenic maximum contaminant level (MCL) of 10 parts per billion (ppb) has a potentially devastating impact on small and rural water systems in the United States (U.S.), particularly since non-transient non-community water systems are also required to comply with the new regulation. Water systems serving fewer than 500 persons (<200 service connections) comprise 72% of the systems impacted by the new arsenic MCL and generally are simple groundwater systems with a single source, storage tank, and hydropneumatic controls. Some of these systems disinfect the water, which represents the most complex process currently utilized by these utilities. Implementing a centralized treatment system at these utilities would be a near-impossible feat, given the lack of operator expertise, financial resources, and technical and management support. In addition, many of the least expensive arsenic treatment processes generate residuals streams and even hazardous wastes. The majority of small systems will have a great deal of difficulty in managing these wastes and complying with environmental disposal regulations. Further, the unit costs per household for a central treatment system utilizing the simplest of arsenic treatment technologies (e.g., throw-away adsorption media) would add $40-$50 per month to the average household water bill for a small system, based on information documented by the research team.

Given the operational, financial, and implementation constraints for very small systems, installation of "under the sink" point of use (POU) devices for arsenic removal may be a more viable and cost effective option. The 1996 Safe Drinking Water Act (SDWA) amendments allow the use of POU and point of entry (POE) treatment devices for compliance with certain MCLs (e.g., inorganics, organics, disinfection by-products [DBPs], radionuclides), but not for microbials. This provision was targeted specifically for small and rural systems that need to remove trace metals and organics using proven technologies. Advantages of a POU system include cost savings for smaller systems, as less than 1% (approximately 0.5 to 1 gallons per person per day) of the total flow generated by a water system would be treated, instead of the 100 to 200 gallons per person per day commonly delivered by public water systems for all uses. Disadvantages of POU systems include management and maintenance of units (at least one unit per home) within costumers' homes, process reliability, public perception, liability if a treatment system fails (due to factors beyond the utility's control), and lack of control of the operation of the units.

OBJECTIVE

The primary objective of this research was to perform an overall feasibility analysis of POU and POE systems for arsenic treatment, and to develop industry-wide recommendations for use of such systems as an alternative to centralized treatment, considering such factors as process reliability, public perception, liability, and regulatory acceptance. The project also assessed water quality criteria that may limit the performance of POU and POE systems and determined at what system sizes POU and POE treatment systems are cost effective, considering management, maintenance, and monitoring expenses. In addition, the project developed guidance for the implementation of a POU and POE treatment program for Arsenic Rule compliance.

CURRENT STATUS OF FEDERAL AND STATE REGULATIONS

To be consistent with the 1996 SDWA amendments, the U.S. Environmental Protection Agency (USEPA) deleted the prohibition on the use of POU devices as compliance technologies in 1998, and has approved centrally managed and controlled POU and POE treatment devices as a means to achieve compliance with the MCLs established in the National Primary Drinking Water Regulations. However, certain criteria have been placed on the use of POUs or POEs for compliance purposes. These criteria include prohibiting the use of a POU device for microbial contaminants; requiring that POU and POE units must be equipped with a warning device (e.g. alarm, light, etc.) that will alert users when the unit is not adequately treating the water; and compliance with independent certification requirements.

Water systems considering implementing a POU or POE treatment program for SDWA compliance must consider state regulations and policies, which vary considerably from state to state. A survey of state regulatory agencies indicated that some states (e.g., Pennsylvania) do not allow the use of POU devices for compliance. Other states allow the use of POU and POE devices for compliance with MCLs for only certain contaminants (e.g., arsenic, radionuclides, VOCs).

RESEARCH APPROACH

A variety of arsenic removal technologies can be used in treatment of drinking water; however, not all of these technologies are suitable for POU or POE devices due to issues including cost and disposal of wastes. Reverse osmosis (RO) and throw-away adsorption media (activated alumina [AA], manganese or iron-enhanced activated alumina [Mn-AA, Fe-AA], and granular iron media [GIM]) are the most viable options for small utilities, and, hence, were the focus of this study.

A twelve-week field screening evaluation was performed to verify the performance of POU and POE treatment units. Four types of adsorptive processes (AA, Fe-AA, Mn-AA, and GIM) and two different membrane types (DOW Filmtec and Osmonics AG) were evaluated. Screening evaluations were performed at six different test sites: Tucson, Ariz. (Metro Water); Stagecoach, Nev.; Sun City West, Ariz.; Unity, Maine; Carson City, Nev.; and Houston, Texas. For the POU units, the first two weeks of testing included a 40-minute on/40-minute off cycle, 16 hours per day, followed by an 8-hour rest period. During the last ten weeks of testing, the operation was designed to mimic actual household operation of the POU units. The POE columns were operated on an intermittent and continuous schedule for the duration of the twelve weeks. Samples were collected three times per week and were analyzed for arsenic, silica, pH, chlorine residual, and conductivity from the POU/POE treated effluent and once per week from the POU/POE influent (raw water). For the POU units, additional parameters such as alkalinity, TDS, sulfate, fluoride, nitrate, iron, manganese, and phosphate were analyzed once per week.

Results from the POU screening evaluation showed that the POU adsorption units performed well and met the manufacturers' criteria of >1,000 gallons treated prior to 10 ppb arsenic breakthrough. Trace levels of iron in the raw water may have enhanced performance of the POU adsorption units at one of the sites. The RO units treated a lower number of gallons before reaching arsenic breakthrough due to the low permeate flux through the system. POU RO treatment failed to remove arsenic (III) at the Unity, Maine site and demonstrated that POU RO is not suitable at this site without pre-oxidation. Based on the results of the POU field screening evaluation, RO, GIM, AA, and Mn-AA were determined to be the most suitable technologies for in-home testing.

All of the POE units demonstrated excellent performance, treating over 50,000 gallons before breakthrough, except the POE Fe-AA unit at Stagecoach, Nev. Varying levels of iron in the raw water at this site may have caused plugging of the column and reduction of the media adsorption capacity. Periodic backwashing of the Fe-AA column would be required for successful POE application at this site.

The most suitable in-home POU units identified in the field screening evaluations were installed in homes at three small communities whose water supply exceeds the 10 ppb arsenic MCL. The three sites selected for the study included the Silverbell Water System, Tucson, Ariz; Stagecoach General Improvement District, Nev.; and Sandy Stream Park, Unity, Maine.

At the Stagecoach and Silverbell sites, RO and Mn-AA POU treatment devices demonstrated overall excellent performance in the in-home testing program, removing arsenic to below detection limits of <0.002 mg/L even after 12 months of continued usage. Only two Mn-AA units had effluent arsenic levels above the detection limits; however, despite arsenic leakage of approximately 5-6 ppb, these units still removed arsenic to levels below the 10 ppb MCL. From the statistical assessment of the in-home POU field testing results, it appears that a conservative POU monitoring frequency for the Silverbell Water System and Stagecoach General Improvement District may be one sample per household per year. However, a reduced monitoring frequency of sampling only one-fourth of the total number of households per year, accompanied by yearly replacement of the membrane or adsorption media may also be acceptable. Optimal monitoring frequencies will vary between communities depending on raw water quality characteristics, seasonal variations in demand, and other site specific issues such as customer acceptance and participation, and state regulations.

Although the Mn-AA cartridge removed As(III) in the initial screening evaluation, the in-home Mn-AA units did not successfully remove As(III). As a result, this study was inconclusive on the removal of As(III) using POU Mn-AA adsorption units.

COST ASSESSMENT

An evaluation of treatment costs for two different scenarios, based on varied monitoring and maintenance criteria was performed. For Scenario 1 (increased monitoring and reduced maintenance), the average market pricing costs for POU adsorption was $26/month while the average market pricing costs for POU RO was $31/month. For Scenario 2 (reduced monitoring and increased maintenance), the average market pricing costs for POU adsorption was $25/month while the average market pricing costs for POU RO was $28/month. The tradeoff between the two options, however, is a high monitoring frequency accompanied by low maintenance operations and vice versa. The costs for both scenarios are similar, whether the service is provided by a vendor on a lease basis or if the utility purchases and maintains the units.

POU treatment is cost-effective when compared to centralized treatment and is viable for small systems. The average breakpoint cost is approximately at 120 connections for POU RO and at approximately 200 connections for POU adsorption.

POE treatment costs are generally higher (39-48%) than centralized treatment costs for systems serving 20 to 300 connections. Therefore, POE treatment may not be considered a feasible option for small water systems, unless a portion of the system's water supply is used for non-residential and non-drinking water purposes, for which treatment to MCL quality is not needed.

POU/POE IMPLEMENTATION BARRIERS AND POTENTIAL SOLUTIONS

Development and implementation of POU or POE treatment programs is complex and involves regulatory, utility, customer, liability, and other issues. Regulatory agencies are faced with addressing issues such as treatment of non-residential taps, 100% household participation, health risks associated with use of non-treated taps, required frequency of POU/POE monitoring and maintenance, disposal of wastes, and administrative burden and costs. Some of these issues may be addressed by requiring ordinances that make use of community water dependent on installation of a POU or POE unit, or by requiring agreements between water systems and customers to allow access to homes to maintain the POU or POE units. Water systems must address issues including household participation and retention in a program, treatment of non-residential taps, and gaining access to homes for installation, monitoring, and maintenance purposes. Customer-related issues play an important role in the development of a successful POU or POE treatment program. These issues include customer acceptance, preference of a POE over POU device, and, in the case of a POU treatment program, use of untreated taps for consumption and cooking purposes. Liability issues such as product liability and liability associated with entering homes for installation, monitoring, and maintenance purposes may be addressed through adequate liability insurance and contracting services with licensed vendors. Customer education and communication are key to overcoming many of the barriers to implementation of a successful POU or POE treatment program. Effective outreach will help in developing a partnership between regulators, water systems, and customers.

CHAPTER 1
INTRODUCTION

ARSENIC TREATMENT FOR SMALL WATER SYSTEMS

The recently promulgated arsenic maximum contaminant level (MCL) of 10 parts per billion (ppb) has a potentially devastating impact on small and rural water systems in the U.S., particularly since non-transient non-community water systems are also required to comply with the new regulation. Systems serving fewer than 500 persons (<200 service connections) generally are simple groundwater systems with a single source, storage tank, and hydropneumatic control system. Of all the systems impacted by the arsenic MCL, 72% of the utilities serve between 25 and 500 people (USEPA, 2001). Some of these systems disinfect the water, which represents the most complex process currently utilized by these utilities. Implementing a centralized treatment system at these utilities would be a near-impossible feat, given the lack of operator expertise, financial resources, and technical and management support. In addition, many of the least expensive arsenic treatment processes generate residuals streams and even hazardous wastes. These small systems will have a great deal of difficulty in managing these wastes and complying with environmental disposal regulations. Further, the unit costs per household for the simplest of arsenic treatment technologies (e.g., throw-away adsorption media) would add $40-$50 per month to the average household water bill for a small system, based on information documented by the research team (NCS, 2000).

Given these operational, financial, and implementation constraints for very small systems, installation of "under the sink" point of use (POU) devices for arsenic removal may be a more viable and cost effective option. The 1996 Safe Drinking Water Act (SDWA) amendments allow the use of POU and point of entry (POE) treatment devices for compliance with certain MCLs (e.g., inorganics, organics, disinfection by-products [DBPs], radionuclides) but not for microbials. This provision was targeted specifically for small and rural systems that need to remove trace metals and organics using proven technologies. Advantages of a POU system include cost savings for smaller systems, as less than 1% (approximately 0.5 to 1 gallons per person per day) of the total flow generated by a water system would be treated, instead of the 100 to 200 gallons per person per day commonly delivered by public water systems for all uses. Disadvantages of POU systems include management and maintenance of units (at least one unit per home) within costumers' homes, process reliability, public perception, liability if a treatment system fails (due to factors beyond the utility's control), and lack of control of the operation of the units.

Many small groundwater systems without any previous treatment experience will be forced to install new, complex arsenic removal systems or multiple POU systems. Maintaining the operational integrity of these arsenic removal systems is essential in achieving any benefit. For small systems serving customers with low and fixed incomes, the cost savings associated with a POU system could be significant.

The U.S. Environmental Agency (EPA) has stated that whole house treatment or point of entry systems are needed where exposure to contaminants (e.g., radon, volatile organic compounds [VOCs]) from routes other than ingestion is the only exposure route of concern for arsenic, whole house treatment would not be necessary (USEPA, 2002). Although POE systems are included as a means for complying with the arsenic rule, the practical implementation issues and economics of POE systems must be considered. By design, POE systems must be capable of treating the entire peak daily flow to a house, the equivalent of a small treatment plant at each house. Space requirements, residuals handling issues, complexity of operations, and costs all make POE treatment unsuitable for arsenic compliance.

Point of use systems offer lower capital costs and may reduce engineering and other fees when compared to centralized treatment. However because of the increased monitoring requirements, the administrative costs will be higher. POU systems may be feasible for small utilities even with higher administrative costs because only a small portion of the total flow (i.e, about 1%) is treated. According to EPA's break point analysis (Preamble to the Jan. 2001 Rule, USEPA, 2001), the POU option is more economical than centralized treatment up to a system size of 72 households or 180 people. However, this analysis does not consider waste disposal costs for the centralized treatment option. Factoring in waste disposal costs would raise the overall costs of centralized treatment, and consequently POU systems might be competitive over a wider range of system sizes.

EPA assumed that systems were more likely to choose POU activated alumina (AA) or reverse osmosis (RO), and POE AA in the Arsenic rule (USEPA, 2001). They did not consider POE ion exchange (IX) or POE RO systems. POE IX systems were not considered because if these resins are not regenerated frequently, sulfate and nitrate can displace the arsenic adsorbed on the resin (chromatographic peaking), causing high levels of arsenic to be discharged in the treated water. POE RO was not considered because of the substantial amounts of brine that would need to be disposed of, and also because of the corrosion problems anticipated due to the low alkalinity waters produced. From a break point analysis, EPA concluded that a POE system would be economical for a system size up to 53 households or 180 people. However, EPA has indicated that POE AA can only be used economically based on raw water pH (less than 7.5) and the capacity of the media.

PERCEPTIONS REGARDING POU AND POE USE

The public has been voluntarily using POU and POE devices for many years for aesthetic reasons such as TDS, taste, odor, or chlorine removal. Currently, there are over 100 manufacturers of under-the-sink devices for water treatment (based on the number of NSF, Water Quality Association, and Underwriters Laboratories drinking water treatment unit product certifications). Sweeping change in utility perception and willingness to partner with the POU/POE industry is necessary for POU and POE systems to become an effective compliance tool. The customer survey components of this study provide some information on the customers' attitudes towards utility (or utility contractors) installing under-the-sink devices. Some of the questions addressed in this study include: Will the savings in treatment costs and the potential health benefits be viewed favorably by the customer? Or, will anti-establishment attitudes prevail where the customer is reluctant to have utility personnel monitor the POU or POE device? Will the majority attitude prevail in a community, or will one or two customers with negative utility perceptions prevent POU or POE implementation and force systems to use centralized arsenic treatment?

Various industry and government associations have stated their positions on the use of POU and POE systems for compliance. EPA's point of view is that it is feasible for a small public water system to own, operate, and maintain POE and POU devices for compliance either directly or through a contract with a qualified vendor (USEPA, 2002). The Association of State Drinking Water Administrators (ASDWA) believes that a transfer of technology is essential to make use of POU/POE systems feasible for compliance (Hamill, 1987). This would involve individuals (customers), regulatory, local utilities, and the home treatment industry and would be used to develop implementation guidelines. ASDWA also feels third party certifications and independent verification of the treatment devices are necessary as these programs are implemented. The American Water Works Association (AWWA) stated that it has no objection to the use of POUs and POEs to improve the aesthetic quality of water; however, AWWA believes that POU/POE devices are not appropriate alternatives or replacements for central treatment of drinking water (Den, 1999). Central treatment of drinking water is the alternative of choice. AWWA believes that central control of POU/POE systems is necessary, although ownership may not be necessary. Rental/lease situations may be appropriate for POU/POE treatment, but the control still has to be under the auspices of the water utility. AWWA pointed out that in addition to beneficial modification to the water quality, there is a potential risk for adverse modification of the water. Consumers need to be educated to understand the potential for adverse modification with POU/POE devices.

OBJECTIVES

The primary objective of this research was to perform an overall feasibility analysis of POU and POE systems for arsenic treatment, and to develop industry-wide recommendations for use of such systems as an alternative to centralized treatment, considering such factors as process reliability, public perception, liability, and regulatory acceptance. The project also assessed water quality criteria that may limit the performance of POU and POE systems and determined at what system sizes POU and POE treatment systems are cost effective, considering management, maintenance, and monitoring expenses. In addition, the project developed guidance for the implementation of POU or POE systems for Arsenic Rule compliance.

ORGANIZATION OF REPORT

Chapter 2 presents a review of several previous studies that investigated the use of POU and POE devices for the removal of arsenic and other contaminants. The experimental details, results, and conclusions from studies in New York, New Mexico, Oregon, Alaska, Nevada, and India and Bangladesh are summarized. A summary of the recently completed AwwaRF project *Conventional and Unconventional Approaches to Water Service Provision,* and some results from the POU field pilot studies are also presented in this chapter.

Chapter 3 summarizes the current status of federal and state regulations pertaining to the use of POU and POE devices, particularly for SDWA compliance. The results of a survey sent to state regulatory agencies to gain information on the states' current, planned, and potential rules and guidance for use of POUs and POEs for compliance and other purposes are presented.

A brief review of arsenic chemistry and treatment and a summary of POU and POE technologies and their applications is presented in Chapter 4. The chapter also presents the results of field screening tests of POU and POE technologies that were performed in Tucson and Sun City West, Ariz.; Stagecoach and Carson City, Nev.; Unity, Maine; and Houston, Texas.

Chapter 5 describes the one-year, in-home POU field testing program that was performed in Tucson, Arizona and Stagecoach, Nevada. Descriptions of the test sites and their water quality are provided, and the testing protocol is summarized. The chapter also presents the results of the POU in-home testing, including information regarding perceptions of the homeowners, obtained from surveys performed after installation of the POUs and at the end of the study.

A cost assessment of POU and POE treatment is presented in Chapter 6. The results of a vendor survey to obtain information on equipment and maintenance costs are summarized. This chapter also presents a comparison of central treatment costs to POU and POE treatment costs.

Chapter 7 presents potential implementation barriers to the use of POUs and POEs for compliance purposes and recommends potential ways to address these constraints. Implementation barriers that are suggested in this chapter address regulatory and statutory acceptance issues and utility, customer, and liability issues.

POU/POE management and implementation guidelines for utilities are provided in Chapter 8. The chapter presents questions that a water system can use to determine whether the system is a candidate for a POU or POE water treatment program. A decision analysis diagram is provided to assist water systems in determining whether POU, POE, or central treatment is the most feasible option for compliance with the Arsenic Rule. In addition, the chapter recommends steps for the implementation of a POU or POE treatment program.

Chapter 9 summarizes the research activities and conclusions from the project. Recommendations for utilities and future research that will supplement the objectives and results of this project are also presented.

CHAPTER 2
REVIEW OF PREVIOUS POU AND POE TREATMENT STUDIES

This review summarizes the results of several previous POU and POE treatment studies performed in the U.S., India, and Bangladesh. Key issues related to the water quality at the study sites are presented, as well as brief descriptions of the study protocols, POU/POE equipment details, cost assessments (if performed in the study), and performance of the POU or POE devices. In addition, conclusions from the studies and lessons learned are provided.

POU AND POE TREATMENT ON LONG ISLAND, N.Y.

A home demonstration study was conducted with POU and POE devices on the North Fork of Long Island, N.Y. in the town of Southold (DeFilippi and Baier, 1987). The groundwater, which is the sole source of water supply in this area through individual private wells, had high levels of agricultural chemicals (nitrates, pesticides, and herbicides). A previous study revealed that construction of a public water supply system would be prohibitively expensive because of the rural nature of the community. The study recommended an investigation into POU/POE devices. A home demonstration project was initiated in September 1985 with 10 manufacturers participating. Eighteen units were installed in homes with various contamination problems. Specific water quality problems were targeted. Since the contaminants in the water varied with location, different types of units were selected for the homes depending on the specific contaminants in the water. For nitrate contamination, a counter-top ion exchange unit was used; for nitrate and chloride problems, an RO POU unit and a carbon filtration device were used; for nitrate and VOC problems, a counter-top distiller was used; for homes with only VOC problems, carbon units were used; for iron and carbofuran contamination, a counter-top filter and carbon POU were used; for homes with only iron contamination, whole house aeration, carbon, and filtration were used; and for manganese contamination, an RO POU unit and a carbon filter were used.

All of the installed units operated satisfactorily and the consumers were pleased with the results. Some sampling results indicated problems which could be traced back to inadequate installation and maintenance. Coliform bacteria were detected in the carbon units and higher numbers of heterotrophic plate count (HPC) bacteria were present in the POU effluent (4 to >6,000 colony forming unit per milliliter [CFU/mL]) than in the POU influent (1 to 1,100 CFU/mL). Organics and metals leaching were also detected due to leaching from the devices and from the associated piping.

Several conclusions were reached from this study, as listed below:

- POU or POE treatment can be used to remove contaminants from drinking water, provided that proper controls are developed for the use of the devices and a central agency (local utility or government) is responsible for implementation and monitoring.
- Equipment installers should be trained, both mechanically and technically, before they install the devices.
- Monitoring of the units must be performed routinely to obtain the efficacy and performance data needed for maintenance schedules.

ARSENIC REMOVAL FROM DRINKING WATER IN SAN YSIDRO, N.M.

Arsenic removal from groundwater high in arsenic and fluoride was studied in San Ysidro, N.M., by Clifford and Lin (1991). San Ysidro is a small, rural village with 70 residences (approximately 200 people) located in north central N.M., approximately 60 miles northwest of Albuquerque. It is bounded on two sides by Native American reservations, and by desert and mountains on the other sides, so options for community water supply are very limited. According to the 1990 census, the average annual income of a family living in San Ysidro is approximately $20,000. Approximately 58% of the families have incomes of less than $10,000 a year. A large portion of the working force commutes to the Albuquerque metropolitan area and thus are not home during the day.

The Village of San Ysidro's water system was built in 1969 and significantly modified in 1983 and 1987. Water is provided by two sources: two shallow infiltration galleries and a 44 foot deep well. The infiltration galleries (constructed in 1969 and 1997) are located near the Jemez River and can produce approximately 27,000 gallons per day (gpd) of water in the winter and approximately 36,000 gpd in the summer. The water supply well was dug in 1982 and can produce 46,000 gpd. The current demand is approximately 35,000 gpd. Because the well and infiltration gallery are strongly influenced by the Jemez River, this water supply has not been adequate for the community during drought conditions, and water has had to be hauled by the NM National Guard in the past.

The concentrations of both As and F in water samples collected from the San Ysidro water system are highly variable. This is likely due to the proximity of the well and infiltration galleries to the Jemez River. Historic river water quality data is limited due to infrequent sampling, but the average As and F concentrations are 42 micrograms per liter (μg/L) and 2.16 milligrams per liter (mg/L), respectively (Clifford and Lin, 1991). The principal source of As is a group of hot springs 15 miles upstream which contain from 800 μg/L to greater than 2,000 μg/L As (Thomson et al., 2002). Water quality data for the Village system is sporadic. Water in the Village well averages approximately 170 μg/L As, 5.2 mg/L F, and 120 mg/L SO_4. Clifford and Lin (1991) reported that As in the Village wells was nearly 100% As(III), while water from the infiltration gallery is about 40% As(III); however, because the water is chlorinated, all of the As in the distribution system was assumed to be As(V). The average pH is 7.0 and levels of silica are high (60 mg/L).

POU treatment was recommended in this study after extensive tests ruled out other treatment options because of poor raw water quality and cost constraints. RO and electrodialysis reversal (EDR) removed sufficient arsenic but were too expensive for the small utility. IX was not feasible because of very short run lengths for arsenic removal and no fluoride removal. AA units had short run lengths prior to arsenic breakthrough (<2000 bed volumes [BVs]) because of high fluoride levels in the raw water. In addition, the regeneration process and sludge disposal problems associated with adsorption media were determined to be too complex for a small community. The Village, like many rural communities in New Mexico, had never been able to attract and retain well-trained operators to run its water system.

These considerations led to the investigation of POU treatment using RO. Initial tests were performed using a Culligan Model No. H-82 RO unit, which was installed at a local community center. The system had a nominal capacity of 8 gpd and contained a 10-micron cartridge filter, a granular activated carbon (GAC) filter, a thin film composite (TFC) RO membrane, a second smaller GAC filter, and a pressurized storage tank. The membrane unit was operated at a low recovery rate of 10-15%. The initial testing results (Clifford and Lin, 1991) indicated that the unit provided good removal of total dissolved solids (TDS) (95%), arsenic (90%), and fluoride (92%).

A preliminary cost analysis was performed for the AA POU, RO POU, and central RO treatment options for this community. Assuming no disposal costs, the AA POU capital costs for a family of four were estimated to be $200-$300, which translates to an increase in the monthly rate of $6 per residence (1988 dollars). This was based on a water usage of 5 gpd for each family and a run length of 500 BVs at a pH of 7.0. The costs for a RO POU unit were calculated at $500/residence for initial installation and an operating cost of $10-$15/month, depending on how long the membranes last. A central RO treatment facility would cost $26.87/month per household, not considering brine disposal (Clifford and Lin, 1991).

After analysis of the results, it was decided to implement POU systems in all of the homes in this community. A public hearing was held in the village to discuss the arsenic problem and to discuss the procedures needed to get the units installed, maintained, and tested during the study period. The village passed an ordinance that made the use of village water contingent upon installation of the RO unit in the home. Each customer also had to sign a permission form to allow the village to install the unit in their home and to allow access to the unit for testing and maintenance. Some customers were reluctant, however they were informed that the village would shut off their water if they did not comply.

All homes were equipped with RO POU systems with Culligan H-82 polyamide spiral wound membrane modules. These units were purchased and installed in May and June 1986. The units were tested every other month for arsenic and fluoride and every three to four months for chloride, iron, and manganese. Coliform bacteria were sampled in 30 of the units. Data from June 1986 to June 1987 indicate that apart from some initial high arsenic levels in the effluent, the units performed well. The raw water arsenic level was 68 ppb while the effluent arsenic level ranged from less than 10 ppb to 20 ppb. The effluent fluoride concentration was less than 1 mg/L. There were reductions in manganese (80%), iron (85%), and TDS (95%), as expected. Routine samples for total coliforms were taken and initially some high positive results were observed. It was suspected that these were due to biogrowth in the carbon filters. Replacement of these filters led to a decrease in the counts.

During the testing portion of the project, difficulty was experienced with collecting samples in the homes. Sampling was performed by an employee of the village and many of the homeowners were at home only on the weekends; sampling 25-30 homes typically took 2.5 days.

In follow-up studies, Thomson and O'Grady (1995) and Thomson et al. (2000) found that when properly operated and maintained, the POU devices are capable of providing very high quality water that effectively removes both arsenic and fluoride. However, the researchers also found that performance of the POU systems in San Ysidro significantly deteriorated over time. As part of the community's SDWA compliance program and also to evaluate the performance of the POU systems, residential and commercial taps were sampled and analyzed for As immediately after the POU systems were installed, and again in 1992, 1993, and 1996. The results showed that about 30% of the POU systems did not remove As to below 10 μg/L.

When they were installed, it was recommended that the POU units be serviced by village personnel once every three months (Rogers, 1990). The recommended maintenance procedures include: (1) inspect the prefilter assembly; (2) replace the prefilter; (3) inspect the carbon postfilter and replace if needed; (4) inspect the RO module housing assembly for cracks or leaks; (5) check all hose connections for leaks; (6) inspect the reservoir tank for cracks or leaks; and (7) restart the unit and inspect it for leaks. A typical routine maintenance check took about 30 minutes to complete. This maintenance has been done for most of the units in San Ysidro, although seldom on a three-month cycle. The water system operator in San Ysidro has no means of measuring the performance of the RO membrane in the POU systems, hence it is replaced only when obvious physical failure is evident.

There are several reasons for infrequent servicing of the POU systems: (1) the residents were not home when the operator arrived to service their unit; (2) some residents prefer not to have outsiders, particularly a village employee, in their home; and (3) some residents do not want the POU treatment system, and avoid having the unit serviced. There is a sentiment among many members of the community, particularly older residents of the Village, that "I've been drinking this water all my life and it's never harmed me." This feeling leads to skepticism regarding the need to treat the water.

It is clear from the results of this study that successful performance of POU systems is strongly dependent on an active operation and maintenance program, testing of the performance of the systems, and regular replacement of expendable POU system components. Therefore, if POU systems are to be used by a public water supply system, it is important that the utility and the regulatory agencies take an active role in the operation and maintenance of the individual customers' systems.

EPA FIELD STUDIES OF POU TREATMENT SYSTEMS FOR ARSENIC REMOVAL

A study was conducted by the EPA to evaluate the effectiveness of IX and AA POE systems and RO POU units in removing arsenic. Tests were conducted in private homes in Eugene, Oregon and Fairbanks, Alaska (Fox and Sorg, 1987). For this study, the treated water from the units was not consumed by the homeowners. The homes all used private wells. The raw water arsenic concentrations at these sites ranged from 50 to 160 ppb. Pilot systems were installed in two homes in each community. Each pilot system consisted of an AA tank (1 cubic foot [ft^3]), an IX tank (1 ft^3), and a RO system. The IX and AA POE systems used cylindrical tanks measuring 46 inches high and 9 inches in diameter. The RO POU systems used were designed to produce 3 to 5 gpd with a source water pressure of 20-100 pounds per square inch (psi) and a recovery of 10%. The average pressure was 40 psi for the installations. The POU units were composed of a cartridge prefilter, cellulose-acetate, spiral-wound membrane, storage tank, and a polishing carbon cartridge filter. The three units were connected in parallel and operated via a system of solenoids and timers. Water was automatically dispensed to the units nine times per day. During each flow interval, one gallon of water was dispensed from the AA and IX units and 0.5 gallons of water from the RO unit. After six months, the protocol was changed to dispense water twice as often to achieve faster arsenic breakthrough. Initially, the AA and IX units performed poorly due to inadequate pretreatment of the media. The problem was rectified with new media, and the units subsequently operated satisfactorily.

The three AA units in Oregon operated for 314 to 406 days at one site and 350 to 427 days at the other site before the MCL (50 ppb at that time) was exceeded. The IX units in Oregon operated from 268 to 314 days at one site and between 350 and 427 days at the other site. The low pressure RO units did not perform well and the MCL was exceeded after 119 and 225 days of operation at the two sites, respectively. At one site, the low pressure RO unit was replaced by high pressure RO, and this unit operated for 350 days without exceeding the MCL.

The AA units in Alaska operated for 680 days at one site and 1,226 days at the other site before the MCL of 50 ppb was exceeded. The IX units in Alaska operated for 1,471 days at one site and 680 days at the other site (with regeneration). Low pressure RO units were used at the Alaska sites and did not perform well; the units failed after operating between 14 and 180 days, depending on the site.

The RO units demonstrated high arsenic removal rates at start up (60-80%), but arsenic removal decreased with time to approximately 50%. A high pressure RO system performed satisfactorily. In their conclusions, the investigators stressed the importance of sampling at start up, insuring proper installation, and monitoring routinely to insure satisfactory performance of the POU units.

STATE OF NEVADA EXPERIENCES WITH POU AND POE TREATMENT

Over the past 15-20 years, the State of Nevada has allowed the use of POU and POE devices for compliance with the fluoride MCL. The State has an MCL of 2 mg/L for fluoride, whereas the Federal limit is 4 mg/L. Several non-transient, non-community systems (trailer parks, industries, military facilities) have attempted to use POE and POU devices for regulatory compliance. In most instances, the units have failed because of a lack of monitoring or use of improper devices. The Town of Gabbs, Nev., represents perhaps the most catastrophic failure of the use of RO POUs for fluoride control. In 1986, homes in this small community, which represents a transient population for a military facility, were equipped with POU devices. Improper monitoring and inefficient RO operation at lower pressures resulted in most homes not complying with the standard. The State eventually sampled the units and made this determination. Other problems included sabotage, vandalism of the units due to a transient population, lack of utility personnel expertise to manage and coordinate sampling and maintenance, and use of other taps for drinking water. Two other experiences with more stable populations, such as trailer parks for retirement communities, demonstrated better results, although POU and POE devices were not utilized as long-term solutions. The POE devices were, in general, felt to be more reliable. The State has indicated that it is open to considering newer POU technologies for long-term arsenic control.

USE OF POE DEVICES FOR ARSENIC REMOVAL IN INDIA AND BANGLADESH

Millions of people in Eastern India and Bangladesh are consuming water with up 2 mg/L (2,000 μg/L) arsenic, and arsenic poisoning has reached epidemic proportions. The sources of supply for these rural communities are "tube wells," or small POE type devices that serve a few homes within a village. Centralized distribution systems are generally present in these villages where tube wells are commonly used. The tube wells are operated by hand pumps that penetrate into shallow portions of the aquifer. The average flow of these units is typically around 2 gpm. To reduce the arsenic levels in these drinking water sources, testing of a dual stage POE unit was conducted in 2000. In the POE unit, the water first passes through an oxidation bed and subsequently an Mn-AA arsenic adsorber. The empty bed contact time (EBCT) of each stage is 5 minutes. The results of one test indicated that approximately 17,000 gallons could be treated prior to breakthrough (Gilles and Odom, 2000). This represented a duration of approximately 60 days and was felt to be a cost-effective treatment option for arsenic control at these rural locations. A more extensive test is being conducted at 2,000 locations using these POE devices.

AwwaRF STUDY COMPARING CONVENTIONAL AND UNCONVENTIONAL APPROACHES FOR DRINKING WATER PROVISION

The recently completed Awwa Research Foundation project, *Conventional and Unconventional Approaches to Water Service Provision* (AwwaRF Project 2761; Raucher et al., 2004), examined a range of unconventional approaches for the provision of drinking water, such as bottled water distribution, POU and POE device installation, neighborhood-scale treatment or polishing of distributed water, water reuse, and dual distribution system applications. These approaches were evaluated based on several criteria, including customer preferences and acceptance, regulatory and statutory constraints and promoters, technical performance and reliability, economic efficiency and cost-effectiveness, legal liabilities, utility capabilities and business models, and lessons learned from other sectors. The goal of the project was to provide an objective, multi-disciplinary evaluation of the costs and benefits of alternative approaches for providing water.

This research project examined these unconventional approaches across two fundamentally different options for utility provision of these services:

- Regulatory compliance. This would include applications in which a utility employs an unconventional approach for all its customers (or all its customers within a given water quality zone).
- Supplemental services. These are approaches in which a utility offers a broader array of water-related services to its customers. Supplemental services are offered that interested consumers can choose to purchase either from the utility or from a utility-affiliated partner.

Although many types of unconventional approaches were evaluated in this research project, the greatest focus was on POU and POE devices. The research included a pilot study of POU devices that were installed in two large urban-area utilities (central facility sites), the Los Angeles Department of Water and Power (LADWP) and Contra Costa Water District, in a controlled bench-test setting. As part of this study, POU devices were also installed in the homes of utility customers under typical in-field operating conditions. POE and neighborhood-scale supplemental treatment case studies were also evaluated in a regulatory compliance context in both small and large utilities. Below is a summary of the results of the LADWP central facility POU pilot study.

The LADWP central facility site was located at the Los Angeles Aqueduct Filtration Plant (LAAFP), which uses ozone and direct filtration for treatment and low doses of iron for coagulation. Chlorine is used as a final disinfectant. The filtered water from the plant, which had an average arsenic level of 4.1 ppb, was used as the influent to the POU units. POU systems installed at the LAAFP included GAC, manganese dioxide-coated activated alumina (Mn-AA) adsorption, prefiltration by GAC and RO, and prefiltration by GAC and ultraviolet (UV) disinfection. The operating criteria for the LAAFP central site included 12 minutes per day of operation (six total cycles consisting of two 1-minute cycles, two 2-minute cycles, and two 3-minute cycles). In addition, the POU units were given two separate 1-week "vacations" during the study. Analytical data from April 2002 through February 2003 showed that the Mn-AA and GAC-RO units removed arsenic to levels below detection. The GAC-UV unit appeared to be the least effective in removing arsenic, with removal of only about 10% of the arsenic in the POU influent. The GAC and

GAC-RO units reached the 500-gallon automatic shut-off point shortly after treating 2,380 BVs. The GAC-UV unit reached 500 gallons in December 2002 but was not shut off and removed until February 2003, after treating approximately 1,093 gallons (5,246 BVs). The Mn-AA unit was replaced in June 2002 because the two-cartridge system shut off after the first 250 gallons, rather than switching to the second cartridge. The second Mn-AA unit had treated 2,500 BVs when it reached the 500 gallon automatic shut-off point in February 2003. Low levels of HPC bacteria were present in the POU influent and the effluent from the GAC-UV unit; however, high HPC levels were observed in the effluent from the Mn-AA, GAC, and GAC-RO units. No coliforms were present in samples from the POU influent or effluent from any of the POU units.

Another test site for LADWP was the Control Gorge Canyon, upstream from the LAAFP in the aqueduct system, where arsenic levels are 30 to 40 ppb. A Mn-AA POU unit was installed in a maintenance facility near a power plant site. The unit was operated manually for about 10 minutes per week; some incidental usage may also have occurred but was not recorded. The Mn-AA unit, which treated approximately 12 gallons in 12 weeks, was replaced with another Mn-AA unit that was expected to have significantly more adsorption capacity in early July 2002. From July 2002 to May 2003, the new Mn-AA unit treated approximately 560 gallons and consistently reduced arsenic to levels below detection. HPC levels were significantly higher in the effluent from the Mn-AA unit than in the POU influent.

The researchers concluded that POU and POE devices – if the appropriate technologies or processes are selected and reliable units are installed – can effectively remove contaminants, including arsenic, and improve the taste and odor of the water. A potential complication to the application of POU or POE devices, however, is the possibility that microbial contaminants may be found in the POU or POE effluent. Periodic disinfection of POU/POE spigots or brief flushing of the spigots before use may help minimize this problem. It was further concluded that the application of POU devices for compliance by small systems with the arsenic rule can result in appreciable cost savings, when compared to the cost of centralized treatment. The researchers also found that consumers like and accept in-home treatment units, and generally encourage their utilities to be involved in some form of supplemental service provision.

USEPA POU TREATMENT STUDY IN GRIMES, CALIF.

In an ongoing USEPA-sponsored demonstration project, researchers from NSF International are evaluating the feasibility of an economically sustainable POU decentralized public water system. This project will determine the different factors involved in using POU treatment for compliance with the arsenic rule in a small community. The city of Grimes in northern California was chosen for this program. Grimes has a population of approximately 300 people in 113 residences and a chlorinated water system served by a well with an arsenic level of about 25 ppb. The town has a school, two restaurants, a store, library, and low-cost mobile homes.

The POU equipment selected for the program was based on adsorptive media with a carbon pre-filter, a built-in water meter, and a shut-off mechanism that can be preset for a specific value. This product was laboratory tested per the protocol in the America National Standards Institute/National Sanitation Foundation (ANSI/NSF) Standard and then field tested on location at Grimes. Based on acceptable performance in these tests, the POU devices, which were provided by Kinetico and consisted of a GAC cartridge followed by two AA cartridges, were installed in most of the houses, the school, and other public locations.

The results of the Grimes pilot study show that the costs of the POU treatment program depend on the technology and pricing scheme that are used. The purchase, installation, and maintenance of the POU devices provided by Kinetico would have cost approximately $270 to $300 per household per year, if retail pricing had been applied. However, the researchers indicated that had the devices and services been provided at cost (rather than at wholesale price), then the annual expense may have been closer to $150 per year.

The POU devices themselves cost $300 retail, but the manufacturer indicated that these units would be made available to utilities at about half that cost. This would indicate that there may be considerable room for a community water system to negotiate a pricing arrangement that would provide appreciable discounts from retail prices. Cost estimates based on utility-favorable pricing provided by the manufacturer indicate that per household annual savings could range from $56 to $71, or about 20%, depending on the interest rate and equipment lifetime used.

The researchers found that regardless of the cost scenarios and technologies used, and even accounting for allowances of $125 to $200 per year per unit for management and reporting, the per household cost is in the range of $17 to $25 per month year (using a 3% interest rate over 10 years). For a community of 100 households, the total system-wide annual costs of compliance using a POU approach would thus be approximately $19,800 to $29,400. If utility-favorable pricing is factored in, the actual costs would most likely be at the lower end of that range, and POU-based compliance is estimated at $20,000 per year for the community water system as a whole. Costs for central treatment in Grimes were estimated at $30,000 per year for annualized capital, annual replacement media, and O&M (USEPA 2004).

SUMMARY OF KEY FINDINGS FROM PREVIOUS POU AND POE TREATMENT STUDIES

Key findings from the previous POU and POE treatment studies that were reviewed include the following:

- POU and POE devices – if the appropriate technologies or processes are selected and reliable units are installed – can effectively remove contaminants, including arsenic, and improve the taste and odor of the water.
- Successful performance of POU systems is strongly dependent on an active operation and maintenance program, performance testing of the systems, and regular replacement of expendable POU system components. If POU systems are to be used by a public water supply system, utility and regulatory agencies must take an active role in the operation and maintenance of the customers' systems.
- Proper controls must be developed for the use of the devices, and a central agency (local utility or government) should be responsible for implementation and monitoring.
- Community water systems may be able to negotiate pricing arrangements with POU manufacturers that would provide appreciable discounts from retail prices.
- Equipment installers should be trained, both mechanically and technically, before they install the devices.
- Improper monitoring and inefficient operation of POU units can result in non-compliance with water quality standards. Other problems observed in previous studies include sabotage, vandalization of the units due to a transient population, lack of

utility personnel expertise to manage and coordinate sampling and maintenance, and use of other taps (by customers) for drinking water.

- With POU units, control of the treatment process is dispersed from a central point to multiple points. Therefore, to maintain the same level of control, more monitoring must take place. Monitoring of the units must be performed routinely to obtain the efficacy and performance data needed for maintenance schedules. Sampling at start up of the units should be performed.
- Sampling costs can be much higher for multiple point testing of POU systems than single point testing of central treatment. Collection of samples can be difficult and time consuming, contributing to higher costs and decreased control over testing.
- Microbial contaminants may be found in the POU or POE effluent. In particular, higher numbers of HPC bacteria may be present in the POU effluent than in the POU influent water. This problem may be reduced by periodic disinfection of POU/POE spigots or brief flushing of the spigots before use.
- Initial and continuing education of the customers must be considered; new customers must be made aware of the system, and existing users must be reminded periodically of their responsibilities.
- In most cases, customers like and accept in-home treatment units and are satisfied with the systems and the water produced by the POU units.
- Application of POU devices for compliance by small systems with the arsenic rule can result in appreciable cost savings for the small systems, when compared to the cost of centralized treatment.

CHAPTER 3
FEDERAL AND STATE REGULATORY SUMMARY

CURRENT STATUS OF FEDERAL REGULATIONS

One of the provisions of the 1996 SDWA Amendments allows the use of POU and POE devices as compliance technologies for small systems. To be consistent with these amendments, the USEPA deleted the prohibition on the use of POU devices as compliance technologies in 1998, and has approved centrally managed POU and POE treatment devices as a means to achieve compliance with the MCLs established in the National Primary Drinking Water Regulations (USEPA, 2002). However, certain restrictions have been placed on the use of POUs or POEs for compliance purposes. These restrictions include:

- POU treatment units may not be used to achieve compliance with an MCL or treatment technique for a microbial contaminant or an indicator of a microbial contaminant. POE devices may be used for these purposes.
- POU and POE units must be owned, controlled, and maintained by the public water system or by a contractor hired by the public water system to ensure proper operation and maintenance of the devices and compliance with MCLs. The system must retain oversight of unit installation, maintenance, and sampling, and is responsible for the quality and quantity of water provided to the community. The system may not delegate responsibility for the operation and maintenance of POU or POE devices installed for compliance purposes to homeowners.
- POU and POE units must be equipped with a warning device (e.g., alarm, light, etc.) that will alert users when the unit is not adequately treating the water. Alternatively, the units may be equipped with an automatic shut-off mechanism.
- Only POU and POE units that have ANSI/NSF standards (if such certification is available for the units) may be used as part of a compliance strategy. If the selected units are covered by these standards, they must have been independently certified according to NSF standards by an accredited laboratory.
- Although not explicitly prohibited under the SDWA, POU devices should not be used to treat for radon or for volatile organic contaminants (VOCs).

ANSI/NSF Standards and Industry Certification

The American National Standards Institute (ANSI) is a private, non-profit organization that administers and coordinates the U.S. private sector voluntary standardization system. ANSI, which does not develop standards itself, helps establish consensus among ANSI Accredited Standards Developers to establish standards.

The National Sanitation Foundation (NSF), an ANSI Accredited Standards Developer, is a private, independent, not-for-profit organization that develops and administers programs relating to public health and the environment in areas of service, research, and education. NSF standards are voluntary consensus standards developed by a joint committee of regulators, users, and industry representatives.

NSF International (NSF), Water Quality Association Gold Seal Program (WQA), and Underwriters Laboratories Inc. (UL) each operate independent, not-for-profit product safety testing and certification organizations. The NSF, WQA, and UL drinking water treatment unit certification programs are each equally accredited by ANSI to test and certify POU and POE treatment devices to the ANSI/NSF standards. These three ANSI-accredited certification programs include testing, retesting, and third-party inspections and audits to verify that each certified product meets all requirements of the relevant standard(s). When a device is certified by NSF, WQA, or UL as meeting an NSF drinking water standard, it means the product does not add anything harmful to the water, the product has passed NSF structural integrity durability testing simulating over 10 years of use with beyond normal household water pressures including water hammer, and that performance claims made by the manufacturer and covered by the standard have been verified as being true.

Current ANSI/NSF standards relating to POUs and POEs include:

- Standard 42: Drinking Water Treatment Units - Aesthetic Effects
- Standard 44: Cation Exchange Water Softeners
- Standard 53: Drinking Water Treatment Units - Health Effects
- Standard 55: Ultraviolet Microbiological Water Treatment Systems
- Standard 58: Reverse Osmosis Drinking Water Treatment Systems
- Standard 62: Drinking Water Distillation Systems

In 1995, NSF found a problem with a listed system that was not providing adequate reduction of arsenic. A new system was installed and was still not found to achieve acceptable reductions. However, retesting of the unit using standard 58 found that the system passed. The protocol at that time called for the system to be challenged with 0.30 mg/L of As (III) and the passing criteria was 0.025 mg/L. Further testing revealed that As (III) exposed to open air in the storage tank was converting to As (V). The carbon filter also helped in converting the As (III) to As (V). As a result, the NSF joint committee on drinking water treatment units decided to withdraw arsenic from Standard 58 and Standard 53. These standards were revised in 2002 to include arsenic and the new arsenic MCL in the testing protocol. The updated standards include a requirement that claims for As (V) reduction only be made on water supplies which have been demonstrated to contain only As (V) species, or on water supplies in which As (III) has been oxidized to As (V) and a residual free chlorine concentration is detectable at the treatment system inlet. The changes also specify the composition and characteristics of challenge water to be used for demonstration testing. As(V), in the form of $Na_2HAsO_4 \cdot 7H_2O$, influent challenge levels should be either 0.30 mg/L or 0.050 mg/L. The maximum allowable As(V) level in the product effluent is 0.010 mg/L. In addition, the standards require that As(V) reduction systems include information on the product packaging that states that the system has been tested and demonstrated to remove As(V) but may not remove other forms of arsenic. Product packaging information must include an arsenic facts sheet and state that the system is to be used on water supplies that contain only As(V) or have a detectable free chlorine residual, and that treatment with chloramine (combined chlorine) is not sufficient to ensure complete conversion of As(III) to As(V). A new NSF standard protocol for testing, verifying, and certifying Drinking Water Treatment Units (DWTU) for all As(III) reduction performance is being currently balloted (June 2004) through the NSF Joint Committee for DWTUs. This protocol, following ballot due process, will be incorporated into ANSI/NSF 53 and 58. This will then allow units to be certified for total arsenic reductions.

USEPA Guidance for POU and POE Treatment Strategies for Compliance

Further information on ANSI/NSF standards and restrictions on the use of POU and POE devices for SDWA compliance can be found in the March 2002 revised final draft of the USEPA *Guidance for Implementing a Point-of-Use or Point-of-Entry Treatment Strategy for Compliance with the Safe Drinking Water Act* (www.epa.gov/safewater/standard/pou.pdf). This guidance outlines some of the issues involved with implementing a successful POU or POE treatment strategy and offers recommendations on how to select, install, maintain, and monitor the equipment, based on the type of water quality problem and the capabilities of the units. The guidance document also summarizes the experiences of more than 20 communities that have installed POU and POE devices to treat a wide variety of contaminants. In addition, the guidance document provides information on costs associated with implementing a POU or POE treatment strategy and potential funding sources. Specific recommendations concerning recalcitrant customers, monitoring, logistics, and use of vendors vs. utility staff for POU/POE installation and maintenance are presented in the sections below.

Recalcitrant Customers

USEPA recommends that water systems educate their customers prior to implementing a POU or POE treatment program using methods such as public hearings, water bill inserts, posters, or notices in print or on radio or television. The POU/POE guidance document includes detailed information regarding public education strategies. To address the issue of recalcitrant customers, USEPA recommends that the system or local government draft an ordinance requiring homeowners to provide the necessary access to the water system. Further, a sample access and maintenance agreement and a sample community ordinance are provided in the guidance document.

Monitoring Requirements

Although USEPA places restrictions on the use of POU and POE devices for compliance purposes, there are currently no specific regulations pertaining to monitoring of those devices. The USEPA POU/POE guidance document states that systems which implement POU/POE treatment strategies must continue to conduct all previously required sampling at wellheads or central points of distribution. Sampling of the effluent from the POU/POE units in each household at least one time during the year in which the units are initially installed is recommended. If the treatment units are successful in reducing the contaminants of concern below the MCL, USEPA recommends that for chronic contaminants (e.g., arsenic), sampling frequency can be reduced to once every three years, with one-third of all units within a system sampled each year on a rotating basis. For acute contaminants (e.g., nitrate), USEPA recommends that sampling be performed more frequently. In addition, the USEPA guidance document suggests that the quality of the water from the POU/POE units can be spot-checked during routine maintenance visits through the use of commercially available field testing kits, electrical conductivity meters (for RO units), and water hardness testing (for cation exchange units). The results of the water quality spot-checks would be used to augment the POU/POE performance data obtained by routine sampling. USEPA guidance also recommends that POU and POE sampling should be coordinated with routine sampling and previously required on-site sampling (e.g., monthly coliform and annual lead and copper sampling) to minimize the number of home visits, thus reducing administrative costs, travel time, and disruption to the customers. Although sample monitoring logs are provided by USEPA, no specific recommendations (other than those previously mentioned) regarding the components or state regulatory agency approval of a sampling plan are provided.

Logistics

Recommendations regarding logistics and administrative tasks required to manage a successful POU or POE treatment program are included the USEPA guidance document. The benefits of hiring a part-time employee to develop schedules for installation, maintenance, and sampling, set up and confirm all appointments, and maintain records of visits and sample results are mentioned. Other specific suggestions include the following:

- Schedule visits to homes near each other for the same day to minimize travel time and maximize productivity.
- Send reminder cards to customers and place confirmation calls to reduce the number of missed appointments.
- Provide homeowners with a record or receipt of the visit to maintain positive customer relations.
- Keep all appointments and notify customers in a timely manner if an appointment must be rescheduled.
- Keep accurate records of all sampling and maintenance visits, work performed, and sample results in a log book or database.

Some guidance is provided regarding the development of an overall management plan for a POU or POE treatment program. USEPA recommends that water system personnel may need to consult with local health or licensing authorities during the development of a management plan to ensure approval of installation, monitoring, and maintenance strategies. However, no specific guidance is provided concerning periodic review of the management plan.

Use of Vendors vs. Utility Staff for POU/POE Installation and Maintenance

To ensure the availability of trained maintenance personnel and eliminate up-front capital costs, USEPA guidance mentions that systems may elect to contract with a vendor to lease POU or POE treatment devices. Another suggested option is to have the vendor maintain the units for a period of time following their initial installation while utility staff are being trained. For utilities that elect to perform installation and maintenance using utility staff (i.e., in-house), USEPA recommends that utility staff participate in training programs designed by States or other organizations (e.g., NSF or Water Quality Association [WQA]), or offered by vendors or equipment manufacturers.

CURRENT STATUS OF STATE REGULATIONS

While USEPA provides guidance for implementing POU or POE treatment strategies, the use of POU or POE devices for compliance purposes is regulated at the state level as the state must approve and support the use of a POU or POE based compliance system. In addition, local communities (e.g., San Ysidro, N.M., as described in Chapter 2) may have specific requirements or ordinances. State regulations and laws vary considerably across the U.S., with some states allowing the use of POUs or POEs for compliance (in some cases, only for specific contaminants) and others forbidding their use.

Requirements for operation and management of a POU/POE treatment strategy may also vary from simple (i.e., the public utility is responsible for all aspects, such as installation, monitoring, maintenance, billing, administration, etc.) to complex, such as in New York. In the state of New York, law S4695-B was created to establish water quality treatment districts for providing POU devices. Each district owns the individual units, installs them, and monitors their performance. Once a water quality treatment district has been formed by local government, it can be operated either as a government-owned district or a franchised operation. In a local government-operated district, district employees have responsibility for purchasing equipment, installation, maintenance, monitoring, performance, billing, and administration. In a franchised operation, the local government still retains ultimate responsibility for the treatment district. The contracted private company or public utility provides all of the home treatment services.

Survey of State Regulatory Agencies

In order to assess the current status of state POU/POE regulations and determine whether states have developed, or are planning to develop, rules, policies, or guidance concerning the use of POU or POE treatment for SDWA compliance, a survey of state regulatory agencies was developed. Questions on the survey (see Appendix A) addressed state experience in regulating POUs or POEs for SDWA compliance; current rules, policies, or guidance in place; planned or proposed activities for SDWA compliance using POU or POE devices; rules, policies, or guidance in place for implementation, reporting, and monitoring criteria; information on systems currently using POU or POE devices for compliance purposes; discharge and disposal issues; and attitudes and perceptions of customers in communities where POU or POE systems have been installed.

The survey was distributed in early February 2003 through the ASDWA to all 50 U.S. state regulatory agencies. The survey was provided in electronic format. Responses were received from 24 states. A tabulation of the responses is presented in Table 3.1 below. Of the 24 states that responded, 15 have policies, rules, or guidance in place for the use of POUs and POEs, and four of these states do not allow the use of POUs in a public water system for compliance purposes. Delaware, Kansas, Missouri, and Washington have systems currently using POUs for SDWA compliance. These systems are small groundwater systems that are utilizing POUs for treatment of nitrate, arsenic, selenium, and/or lead. Only three states (New York, Pennsylvania, and Wisconsin) have systems using POEs for compliance purposes (for VOCs and microbial contaminants). Nine states, including five in the eastern part of the U.S., indicated that POUs or POEs are allowed for treatment of arsenic. Several states (Alaska, Arizona, Illinois, Massachusetts, New York, Virginia, Washington, Wisconsin) have planned or proposed activities for SDWA compliance using POU/POE systems. These activities include the following:

- Use of POU and/or POE systems for compliance and public health protection for arsenic, radionuclides, and nitrate.
- Development of guidance for the state to determine the acceptability of POU and POE systems.
- Development of guidance on the use of POE units by public water systems.

Table 3.1
Summary of survey responses from state regulatory agencies

State	1. State agency with primacy authority for POU/POE rules, policy & guidance	2. POU/POE rules, policy, or guidance in place for implementation, reporting?	3. POU/POE rules, policy, or guidance in place for monitoring criteria?	4. State experience in regulating POU/POEs for SDWA compliance	
				Limitations	Potential solutions
Alaska	Dept. of Environ. Conservation	None in place	None in place	Not available	Not available
Arizona	Dept. of Environ. Quality	POU & POE rules in place, policies & guidance in process	POU & POE rules in place, policies & guidance in process	Not available	Not available
California	Dept. of Health Services	POE requirements for irrigation districts	POE requirements for irrigation districts	POE operating cost for small system	None
Delaware	Dept. of Health & Social Services	POU policy under development	POU policy being developed	Use of POUs limited to single service connection	Not available
Florida	Dept. of Environ. Protection	POU and POE rules	POU and POE rules	No experience	No experience
Idaho	Dept. of Environ. Protection	POU guidance, limited POE guidance	POU guidance in place, limited POE guidance	Achieving full participation by individual customers	National regs/guidance to require full participation
Illinois	Rules - IL Pollution Control Board	POU rules for emergency situations	POE rules in place	Just getting started, will adjust as needed during development	Hardness as indicator of radium content
	Policy - Illinois EPA	POE rules allow installation POE policy & guidelines being developed	POE policy & guidelines under development		
Indiana	Dept. of Environ. Management	None in place	None in place	Not available	Not available
Kansas	Dept. of Health & Environment	None in place	None in place	No answer provided	No answer provided
Maine	Maine Drinking Water Program	None in place	None in place	Not available	Not available
Massachusetts	Dept. of Environ. Protection	POU & POE rules (310 CMR 22.23) POU & POE policies under development	POU & POE rules in place, POU/POE policies being developed	No experience to date	No experience to date
Michigan	Dept. of Environ. Quality Water Div.	None in place	None in place	No answer provided	No answer provided
Missouri	Dept. of Natural Resources, Public Drinking Water Program	None in place	None in place	No answer provided	POUs for PB removal in school drinking fountains
Nevada	State Health Div, Bureau of Health Protection Services	None in place	None in place	No answer provided	No answer provided
New York	State Dept. of Health	POE rules in place, POE policy & guidance under development	POE policy & guidelines under development	Insuring continued O&M of the units	Require regular reporting as part of routine monitoring
North Dakota	Department of Health	None in place	None in place	No answer provided	No answer provided
Pennsylvania	Dept. of Environ. Protection, Bureau of Water Supply & Wastewater Mgt.	POU & POE rules in place	POE rules in place	POU only treats at single tap & not whole house	Restrict POUs to temp. use, restrict POEs to v. small systems
Rhode Island	Department of Health, Office of Drinking Water Quality	POU & POE rules in place	POE rules in place	POUs not allowed for compliance	No answer provided
S. Carolina	Dept. of Health & Environ. Control	POU & POE rules in place	Not available	No answer provided	No answer provided
Utah	Dept. of Env. Quality, Div. of DW	None in place	None in place	Not available	Not available
Vermont	State Water Supply Division	POU policy (case by case basis)	None in place	Only 1 potable tap	No answer provided
Virginia	Department of Health	POU & POE rules in place	No answer provided	No answer provided	No answer provided
Washington	State Dept. of Health	POU & POE policies in place POU & POE guidance under development	None in place (but under development)	Securing access to homes, 100% of connections must be treated	Restrict units to non-community settings; avoid monitoring at POU/POE taps
Wisconsin	Dept. of Natural Resources	POU rules in place	None in place	Not available	Not available

(Continued)

Table 3.1 (Continued)

State	5. Guidance/procedures in place if segment of community does not want to install POUs?	6. POUs or POEs allowed for which contaminants?	7. What system sizes can use POUs by regulation or policy?	8. What size systems are currently using POU devices?	9. Info. on systems currently using POUs for compliance purposes
Alaska	None	None	None	None	Not available
Arizona	In process of development	Radionuclides, As, Cr, VOCs	All	None	Not available
California	All must participate, water system responsible for resolving problem	Radionuclides, As, Cr, turbidity, microbials, VOCs	All	None	Pilot study in comm of 200, treating for As using activated alumina
Delaware	Not available	Nitrates	25-100	25-100	GW systems serving <100, nitrates, RO & Ion Exchange
Florida	None	Radionuclides, As	25-100, 101-1000	None	None used
Idaho	None at present	As, Cr, nitrate, F, Pb, Ba, Be, Cd, Cu, Se, thallium, cadmium	<25, 25-100	None	2 small systems - <100 - have explored POUs but not implemented
Illinois	Need to have 100% compliance	Only radium at this time	None	None	None used
Indiana	None in place	May only be used temporarily	None	None	None
Kansas	None	None	No answer provided	25-100	57 connections, GW, RO for Se & As
Maine	Not available	As under consideration	None	None	Not available
Massachusetts	No POU units approved for homes in a PWS	As, Cr, compliance with other MCLs if ANSI/NSF certified	No answer provided	No answer provided	No experience with POUs for compliance
Michigan	No answer provided	No answer provided	No answer provided	None	No answer provided
Missouri	None	Lead, POU/POE use considered on a case by case basis	None	501-3300 (515)	Elem school, GWw/Pb, cartridge POUs on drinking fountains
Nevada	None	No answer provided	No answer provided	None	No answer provided
New York	None	Radionuclides, As, Cr, turbidity, microbials, VOCs for POEs	None	None	No POU
North Dakota	No answer provided	No answer provided	No answer provided	None	No answer provided
Pennsylvania	POUs not acceptable for compliance short-term use only accepted	Radionuclides, As, Cr, turbidity, microbials, VOCs for POEs	None	None	POUs not accepted for compliance short-term use only accepted
Rhode Island	None, all must be protected	No list of approved applications, case by case basis	None	None	No answer provided
S. Carolina	No answer provided	No answer provided	No answer provided	No answer provided	No answer provided
Utah	Not available	No answer provided	None	None	Not available
Vermont	None, this would be a major problem	Radionuclides, As, microbials, VOCs - Yes; turbidity - maybe	None (case by case basis)	None	Not available
Virginia	No answer provided	No answer provided	No answer provided	No answer provided	No answer provided
Washington	100% of community must be covered, policy being developed to prohibit POU/POEs for community WSs	Radionuclides, Cr, As, VOCs non-community only	All	25-100, 101-500	small GW systems with POUs for nitrate treatment will not be allowed in future
Wisconsin	Not available	Radionuclides, POEs - some TCR and nitrate	All	None	None currently installed in community systems

(Continued)

Table 3.1 (Continued)

State	10. Info. on system currently using POEs for compliance purposes	11. Experience/policies re. to WQ degradation due to membrane fouling, microbial degradation, loss of adsorptive capacity, other?	12. For communities using POU systems, estimate % that discharge wastewater to sewer, and % that discharge to septic systems	13. Concerns regarding wastewater from RO POUs and its disposal?	14. Attitude & perceptions of consumers in community where systems have been installed?
Alaska	Not available	Not available	Not available	Yes, will be a concern	Not available
Arizona	Not available	No answer provided	Not available	Not available	Good
California	Not available	Not available	Not available	None identified	Excellent, not clear if attitude will remain as positive once study is over & water system takes over
Delaware	Not available	Not available	Not available	No	No answer provided
Florida	None used	None identified	Not available	None identified	Not available
Idaho	Not available	None identified	100% septic systems	Yes	Average, unknown
Illinois	None so far, but have several GW supplies <500 that will use IEx for radium removal	Not available	Not available	No answer provided	No answer provided
Indiana	None	Not available	None	No answer provided	Unknown
Kansas	No answer provided	No answer provided	100% septic systems	No	Too early to tell
Maine	Not available	No answer provided	No answer provided	No answer provided	No answer provided
Massachusetts	Not available	Not available	Not available	Yes, brine disp regulated	No answer provided
Michigan	No answer provided	No answer provided	No answer provided	No answer provided	No answer provided
Missouri	Not available	No known problems	Not known	Not available	Excellent
Nevada	No answer provided	No answer provided	No answer provided	No answer provided	Unknown
New York	Limited use of POEs for SW & GW w/ VOCs & private wells	Yes - all, extensive experience w/ use of POEs for VOCs at wells	Not known	None in place at PWSs	Good, average
North Dakota	No answer provided	No answer provided	No answer provided	No answer provided	No answer provided
Pennsylvania	Small GW systems using UV POEs for disinfection	No answer provided	No answer provided	No answer provided	Unknown
Rhode Island	No answer provided	No answer provided	No answer provided	No answer provided	No answer provided
S. Carolina	No answer provided	No answer provided	No answer provided	No answer provided	No answer provided
Utah	Not available	No answer provided	Not available	Not available	Unknown
Vermont	Not available	No answer provided	No answer provided	Yes, would be regulated & be a problem for leachfields	No answer provided
Virginia	No answer provided	No answer provided	No answer provided	No answer provided	No answer provided
Washington	Not aware of any, some POEs may be used for single connection systems	Membrane fouling, policy requiring alarms under development	100% septic systems	None. Small amount of WW, all on septic systems	Unknown
Wisconsin	Small non-community GW systems for TCR compliance	No	Not available	Yes	Unknown

(Continued)

Table 3.1 (Continued)

State	15. Who is responsible for O&M of POU/POE devices?	16. What % of POU/POEs comply with SDWA MCLs?	17. What % of POU/POEs meet monitoring requirements?	18. Planned or proposed activities for SDWA compliance using POU/POE systems?	19. Related local or county regulations in state and contacts?
Alaska	Not available	Not available	Not available	Compliance & PH protection for As, radionuclides, nitrates	Not available
Arizona	Utility or 3rd party contractor	Not available	Not available	Currently monitoring 2 POU pilot projects	No
California	Equipment vendor (for pilot study)	Not available	Not available	Only to advise small WSs of this potential option for achieving compliance	No
Delaware	Utility	>90%	>90%	Waiting on EPA guidance	Not available
Florida	Not available	Not available	Not available	None	Not available
Idaho	Utility, 3rd party contractor, equipment vendor	Not available	Not available	None planned, awaiting further experience & interest	No answer provided
Illinois	Not available	Not available	Not available	Radionuclides using POE, may expand to other MCLs as experience is gained	No answer provided
Indiana	Not available	Not available	Not available	None, unless for emergencies	None
Kansas	City contract w/ vendor	No answer	No answer	None at present	Not aware of any
Maine	No answer provided	No answer	No answer	Under consideration for As	Not aware of any
Massachusetts	No answer provided	No answer	No answer	Developing guidance to det. DEP acceptance of POU/POEs	None
Michigan	No answer provided	No answer	No answer	May allow for contaminants reg. by Fed rule, but not microbials	No answer provided
Missouri	Utility	>90%	>90%	May allow POU/POEs in existing PWs on case by case basis	Not available
Nevada	No answer provided	No answer	No answer	Subject to reg. development	No
New York	Utility (resp.), 3rd party contr. (manage)	>90%	>90%	Will provide guidance on use of POEs by PWs	State regs - WQ treatment districts, private well guidance for activated carbon units
North Dakota	No answer provided	No answer	No answer	Will review apps to use POU & POEs on a case by case basis	None
Pennsylvania	Utility	>90%(POE)	>90%(POE)	None at this time	Not known
Rhode Island	POE-water supplier, no POUs allowed	No answer	No answer	No answer provided	None
S. Carolina	No answer provided	No answer	No answer	No answer provided	No answer provided
Utah	No answer provided	No answer	No answer	None for now	None
Vermont	No answer provided	No answer	No answer	None	None
Virginia	No answer provided	No answer	No answer	Considering policy that allows devices if utility owns & maintains	No answer provided
Washington	50% utility, 50% 3rd party contractor	>90%	No answer	Policy being developed - only allows devices in non-community, POU only for chronic contam, POE acute & chronic	Local plumbing codes?
Wisconsin	3rd party contractor; equipment vendor	>90%	75-90%	One system (>3300) investigating POE for radionuclide compliance	POE only allowed for non-community systems

- Proposal of a policy that will allow POU and POE systems if they are owned and maintained by the water system.
- Development of a policy that will only allow use of the devices in non-community systems - POUs only for chronic contaminants and POEs for acute and chronic contaminants.
- Allowance of POUs and POEs on a case-by-case basis.
- Use of data from POU pilot projects to develop policies and guidelines for use of POUs for compliance.

Completed surveys were received from approximately one-half of the states that are the most affected by arsenic contamination in their groundwaters. Arizona, California, Maine, Michigan, Nevada, Utah, and Washington provided a wide range of responses. Only Arizona, California, and Washington indicated that they have any type of POU or POE rules or policies in place and that POUs or POEs are allowed for arsenic removal. California and Washington were also the only states of the above seven that indicated that POUs are being used for compliance purposes by small systems. The California system, however, is using POUs as part of a pilot study. Four of the seven states responded that they have planned or proposed activities related to SDWA compliance using POU/POE systems. Arizona is currently monitoring two POU (RO) pilot projects and will use the data to develop policies and guidelines for the use of POUs for compliance. Maine is considering allowing POU/POEs for arsenic removal in non-community, non-transient systems. Michigan may allow use of POU/POEs for contaminants, including arsenic, regulated by Federal rule. In Washington, use of POUs by small community groundwater systems will not be allowed when new policies currently under development are implemented.

State Regulatory Documents

In addition to completing the survey, the state regulatory agencies were asked to provide policy memoranda, guidance documents, regulations, or other related documents to supplement the survey. Several state agencies provided the following supporting documents (details are presented in Appendix B).

- California Department of Health Services (CADHS) - Requirements for Point of Entry Systems on Constructed Conveyance Water Systems (Irrigation Districts). This document addresses monitoring and oversight requirements and responsibilities, approval of existing POE devices, new POE device installations, monitoring and maintenance criteria, operator qualifications, customer education and operational agreement, annual reporting requirements, and provides a list of Department approved POE devices (CADHS, 2000).
- Idaho Department of Environmental Quality (IDEQ) - Guidelines for Approving Managed Point-of-Use Treatment in Idaho Public Water Systems. These guidelines provide information for submitting and reviewing POU applications and address technical and management issues (e.g., health risks, affordability, operation and maintenance, microbial concerns, compliance monitoring, advantages for small systems, and reporting and record keeping considerations) (IDEQ, 2000).
- Pennsylvania Department of Environmental Protection (PADEP) - sections pertaining to POU and POE treatment devices from Chapter 10.9, Safe Drinking Water

POE/POU Regulations. These sections of the regulations address general monitoring and sampling requirements, public notification requirements, public water system construction permits, acceptable design, approval of POE devices, and control of POE devices (PADEP, 2003).

SUMMARY OF FEDERAL AND STATE REGULATIONS

Although the USEPA has approved the use of POU and POE treatment devices to achieve compliance with MCLs established in the National Primary Drinking Water Regulations, certain restrictions have been placed on the use of POUs or POEs for compliance purposes. These restrictions pertain to the ownership, control, and maintenance of the devices, coverage by and ability to meet ANSI/NSF standards, a requirement that POU and POE devices have warning or shut-off mechanisms, and treatment of microbial and volatile contaminants and radon. USEPA provides helpful guidance for implementing a POU or POE treatment strategy for compliance with the SDWA.

Water systems considering implementing POU and POE treatment devices for SDWA compliance must consider state regulations and policies, which vary considerably from state to state. (Local ordinances and requirements must also be considered, but these are developed within the scope of state regulations.) Some states (e.g., Pennsylvania) do not allow the use of POU devices for compliance. Other states allow the use of POU and POE devices for compliance with MCLs for certain contaminants (e.g., arsenic, radionuclides, VOCs). A survey of state regulatory agencies indicated that several states have plans to develop policies and guidance for using POU/POE systems for SDWA compliance.

CHAPTER 4
POU AND POE TREATMENT TECHNOLOGIES AND PERFORMANCE TESTING

ARSENIC CHEMISTRY AND REMOVAL

Arsenic in water is normally present in two oxidation states - arsenite (As^{+3}) in anaerobic/anoxic (reduced) systems and arsenate (As^{+5}) in aerobic (oxidized) systems. There are two primary reasons why the oxidation state of arsenic is important: 1) As^{+3} is a greater health concern than As^{+5} (Korte and Fernando, 1991; Knowles and Benson, 1983) and 2) treatment process efficiency is lower for As^{+3} than for As^{+5} (Frank and Clifford, 1986; Hecht et. al., 1992).

Due to differences in ionic charge of arsenate and arsenite particles in the pH 6-9 range, the neutrally charged arsenite compound (H_3AsO_3) is difficult to remove when compared to the divalent ($HAsO_4^{-2}$) and monovalent arsenate ($H_2AsO_4^{-}$) compounds. Since most of the treatment processes that are used to remove arsenic cannot reduce As^{+3} to low levels, preoxidation is required to convert As^{+3} to As^{+5} prior to removal. A common method of converting As^{+3} to As^{+5} is by the addition of an oxidizing agent such as chlorine. EPA recommends the use of POU devices on waters that are chlorinated so that most of the arsenic is in the +5 oxidation state. If the waters are not chlorinated, other alternatives may be needed to convert As^{+3} to As^{+5}.

POU AND POE TECHNOLOGIES AND THEIR APPLICATIONS

A variety of arsenic removal technologies can be used in treatment of drinking water; however, not all of these technologies are suitable for POU or POE devices due to issues including cost and disposal of wastes. Three treatment systems potentially suitable for POU and POE devices are: 1) IX; 2) throw-away adsorption media such as AA, Mn-AA, iron-enhanced activated alumina (Fe-AA), granular iron media (e.g., granular ferric hydroxide [GFH]); and 3) RO (Cadmus Group, 1998). RO and throw-away adsorption media are the most viable options for small utilities, since they don't involve complex regeneration activities; these technologies are therefore the focus of this study. A description of these treatment technologies is presented below.

Adsorption Processes

Adsorption refers to the accumulation of material at the interference between two phases (e.g., water and solid media). As water containing ionized arsenic passes through an adsorption column, arsenic attaches to the solid media. Several types of adsorbents that were tested for POU and POE devices under this project are discussed below.

Activated Alumina

AA is a granular aluminum trioxide media that effectively adsorbs arsenic (As^{+5}) (Bellack, 1971; Gupta and Chen, 1978). The product is NSF certified and has been previously used in a wide range of applications. Previous studies conducted in Albuquerque, N.M.; Fallon, Nev.; and with Salt and Verde River waters, Ariz., with EBCTs between 1.5 and 10 minutes indicated that the optimum EBCT for AA is 5 minutes. However, for POU type applications, lower

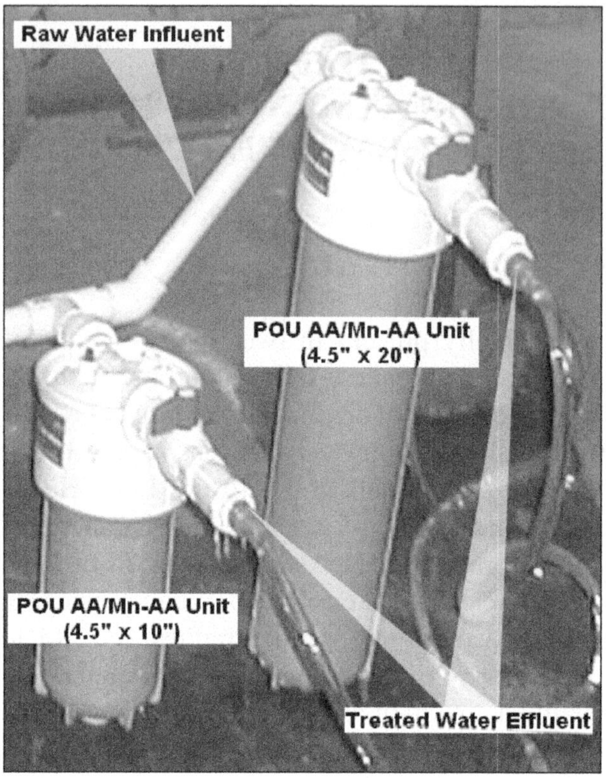

Figure 4.1 POU AA/Mn-AA unit

EBCTs of 1.7 to 3.4 minutes may be applied. AA is regenerable, but significant loss of capacity (e.g., 10% or more) can occur in each cycle (NCS, 2001a). Throw-away media is viable for POU/POE type applications, as it reduces media regeneration and residuals handling costs. Water quality parameters that may impact arsenic removal are silica >20 mg/L, fluoride >2 mg/L, iron >0.3 mg/L, manganese >0.05 mg/L, and sulfides >0.03 mg/L.

The POU AA cartridge is available in sizes of 4.5" × 10" and 4.5" × 20" with EBCTs of 1.7 and 3.4 minutes, respectively, at the typical flow rate of 0.4 gpm. A typical POU AA unit in both sizes is shown in Figure 4.1.

The vendor-based AA media tested during this study was Apyron's Aquabind. The manufacturer product specifications for this media are presented in Appendix C. POU AA was tested at Metro Water, Tucson, Ariz., and Sun City West, Ariz.

Iron Enhanced AA (Fe-AA)

Fe-AA is an enhanced aluminum-based media specifically manufactured for the removal of arsenic (As^{+5}) in drinking water. The media contains additives such as iron to enhance its adsorption capacity. The product is NSF certified and has been previously used in a wide range of applications. The media can be used on a throw-away basis, and the spent media can be disposed as a non-hazardous solid waste. Water quality parameters that may impact arsenic removal are silica >20 mg/L, fluoride >2 mg/L, iron >0.3 mg/L, manganese >0.05 mg/L, and sulfides >0.03 mg/L. Previous testing conducted in Phoenix, Ariz. on an influent with 14 ppb arsenic showed that with

Figure 4.2 POE Fe-AA and GFH units

an EBCT of 5 minutes and with pH adjustment to less than 7.0, Fe-AA treated approximately 18,500 BVs (64 continuous operational days) of water before arsenic breakthrough occurred (NCS, 2001a). The optimal EBCT for Fe-AA treatment is 5 minutes.

The POE Fe-AA columns tested during this project were 8" and 12" diameter fiberglass tanks with EBCTs of 2.5 and 5.0 minutes, respectively. A typical 12" diameter POE Fe-AA unit is shown in Figure 4.2.

The Fe-AA media tested during this study was Alcan's FS-50. The manufacturer product specifications for the Fe-AA media are presented in Appendix C. POE Fe-AA was tested at Metro Water, Tucson, Ariz.; Stagecoach, Nev.; Sun City West, Ariz.; and Houston, Texas.

Manganese Coated AA (Mn-AA)

Mn-AA is an enhanced aluminum-based media specifically manufactured for the removal of arsenic (As^{+5}) in drinking water. The media contains additives such as manganese to enhance its adsorption capacity. The product has been previously used in a wide range of applications. The media can be used on a throw-away basis and the spent media can be disposed as a non-hazardous solid waste. Water quality parameters that may impact arsenic removal are silica >20 mg/L, fluoride >2 mg/L, iron >0.3 mg/L, manganese >0.05 mg/L, and sulfides >0.03 mg/L. The optimal EBCT for Mn-AA for POU type applications is similar to AA and is approximately 1.7 to 3.4 minutes.

The POU Mn-AA cartridge is available in sizes of 4.5" × 10" and 4.5" × 20" with EBCTs of 1.7 and 3.4 minutes, respectively. A typical POU Mn-AA unit is similar to AA and is shown in Figure 4.1.

The Mn-AA media tested during this study was Apyron's Aquabind. The manufacturer product specifications for the Mn-AA media are presented in Appendix C. POU Mn-AA was tested at Metro Water, Tucson, Ariz.; Sun City West, Ariz.; Unity, Maine; and Carson City, Nev.

Granular Ferric Hydroxide

Granular ferric hydroxide (GFH) is an arsenic removal technology developed in the mid-1980s by the Technical University of Berlin, Germany, Department of Water Quality Control and is manufactured commercially by GEH Wasserchemie in Germany. GFH is a fixed bed adsorption process used for removing As^{+5}. The product is NSF certified and is considered highly viable for arsenic removal in small systems if throw-away media is considered. Water quality parameters that may impact arsenic removal are phosphorus >0.2 mg/L, iron >0.3 mg/L, and manganese >0.05 mg/L.

The adsorptive capacity of GFH is several times greater than AA, as confirmed in studies conducted in Phoenix, Ariz. (NCS, 2001a). The results of these tests showed that 50,000 to 80,000 BVs were treated prior to arsenic breakthrough. Previous studies conducted in Phoenix, Ariz. on an influent with 8.9 ppb arsenic showed that with an EBCT of 2.5 minutes and an ambient pH of 7.9, GFH treated approximately 89,000 BVs (155 continuous days of operation) of water before arsenic breakthrough occurred (NCS, 2001a). Results from pilot scale studies indicated that an EBCT of 2.5 minutes is required for effective arsenic removal. However, for POU type applications, EBCTs of approximately 1.7 to 3.4 minutes may be applicable.

The POE GFH columns used during this project are 8" and 12" diameter fiberglass tanks with an EBCT of 2.5 minutes. A typical 12" diameter POE GFH unit is shown in Figure 4.2.

During this study, GFH media was tested at Metro Water, Tucson, Ariz.; Stagecoach, Nev.; Sun City West, Ariz.; Carson City, Nev.; and Houston, Texas. GFH product specifications (provided by US Filter) are presented in Appendix C. Other types of iron media, similar to GFH, have also been introduced to the market since the time these tests were conducted. These include Bayoxide E-33 and As-X; all these perform similarly, with some site specific variations. Collectively, these media can be generically referred to as granular iron media (GIM).

Reverse Osmosis

RO is a pressure driven process where water enters one or more modules which contain membranes. All membrane processes separate feed water into two streams. The concentrate stream contains the substances removed from the feed water after being rejected by the membrane barrier. RO is a low flux membrane process with >95% rejection rates for arsenic. RO removes arsenic and other dissolved constituents and produces water with a low TDS concentration and high quality (NCS, 2001b). In areas where additional contaminants are present in the source water, RO may be the desired treatment system (Fox, 1989). RO performance is sensitive to the concentration of suspended solids, turbidity, iron, calcium, magnesium, and other dissolved constituents. The primary parameters of concern for POU membranes are sulfates, hardness, barium, and NOM, assuming sufficient pre-filtration is provided. The type of RO membranes used in POU type applications are single element membranes, which can produce waste rates of >80%

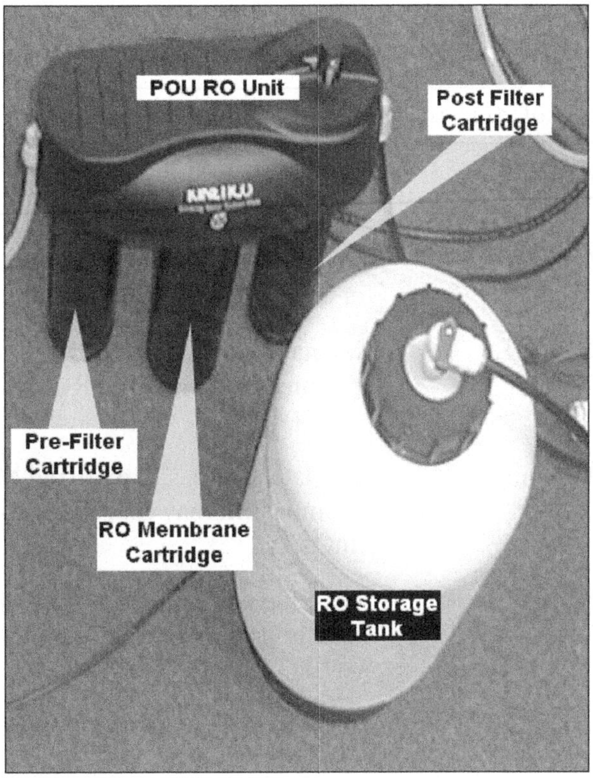

Figure 4.3 POU RO unit

of the influent flow (four gallons wasted for each gallon treated.). These high waste rejection rates may be a concern in water scarce regions. A typical POU RO unit with a pre-filter and a post-filter is shown in Figure 4.3.

The typical POU RO unit consists of a pre-filter cartridge, an RO membrane, a holding tank, a faucet, and a post-filter cartridge. Feed water is input to the pre-filter. After pre-filtering, water flows to the RO membrane. As treated water is produced, it is stored in the holding tank under pressure (bladder principle). The more water that is stored, the higher the tank pressure. From the holding tank, water flows to the post-filter and then is delivered to a faucet. The pre-filter cartridge is a carbon/sediment cartridge for removal of sediment and chlorine prior to introduction to the membrane element. The post-filter cartridge is a GAC cartridge used for improving tastes and odors. The two types of RO membranes tested during this study were the DOW Filmtec membrane and the Osmonics AG membrane. Although Kinetico also manufacturers and supplies RO membranes, these were not used during this study. The typical dimensions of these membranes are 13.13" × 16.37" × 5.18". The detailed specifications of the RO membrane are presented in Appendix C. The DOW Filmtec membrane was tested at Metro Water, Tucson, Ariz. and the Osmonics AG membrane was tested at Sun City West, Ariz.; and Unity, Maine.

WATER QUALITY IMPACTS ON ARSENIC TREATMENT

Generally, arsenic removal in central treatment systems is affected by the presence of other constituents such as silica, phosphorus, pH, fluoride, sulfate, chloride, TDS, sulfides, iron,

Table 4.1
Water quality parameters affecting POU treatment systems

Parameter	Level of concern	Impact on arsenic removal
Silica	>50 mg/L	Affects adsorption processes by reducing As removal capacity. Also fouls RO membrane
Fluoride	>2 mg/L	Affects adsorption and membrane processes as it competes for adsorption sites along with arsenic
Iron	>0.3 mg/L	Arsenate attaches to oxidized iron and plugs adsorption columns. Iron may foul RO membrane by formation of $Fe(OH)_3$
Manganese	>0.05 mg/L	Arsenate attaches to oxidized Mn and plugs adsorption columns. Mn may foul RO membrane by formation of MnO_2.
Sulfides	>0.2 mg/L	May plug adsorption column and cause reduced capacity. Usually present in reduced waters (where As is present at As(III)
Calcium, magnesium, barium, sulfate, carbonate	>saturation potential	Cause inorganic scaling on membrane surface
NOM		Causes membrane fouling, reduced flux rates and reduced rejection rates

and manganese. Parameters that may directly affect processes (e.g., may lead to fouling of membranes) are: iron, manganese, calcium, magnesium, barium, and NOM. Water quality parameters that affect arsenic removal efficiency are pH and alkalinity. Parameters that compete with arsenic for removal are silica, vanadium, phosphorus, fluoride, sulfate, chloride, TDS, and nitrate. For example, silica levels greater than 50 mg/L and fluoride levels greater than 2 mg/L affect adsorption using alumina based sorbents, and pH values greater than 8.5 and phosphorus levels greater than 0.5 mg/L affect iron based adsorption POU units.

However, the expectation for a POU or POE treatment system with respect to performance is far less when compared to a central system. For a central adsorption system to be cost-effective, fouling issues need to be minimized and generally, over 50,000 BVs need to be treated prior to arsenic breakthrough. However, for a POU system, where only a small fraction of the flow is treated, a higher level of fouling can be tolerated and the expectation for system performance is around 1,000 BVs. However, if the interfering parameters are present in extremely high levels, the POU system may not perform adequately to treat even 1,000 BVs. The primary parameters of concern for POU adsorption devices are silica, fluoride, iron, manganese, and sulfides. The primary parameters of concern for POU membrane systems are sulfates, hardness, barium, and NOM. The impacts of each of these parameters on arsenic removal are summarized in Table 4.1.

Table 4.2
POU/POE unit specifications

Location	Technologies
Tucson, Ariz.	POU RO: DOW Filmtec membrane POU: 4.5" × 20" Aquabind EP (AA) POE: 12" column GFH POE: 12" column Alcan FS-50 (Fe-AA)
Stagecoach, Nev.	POE: 12" column GFH POE: 12" column Alcan FS-50 (Fe-AA)
Sun City West, Ariz.	POU RO: Osmonics AG membrane POU: 4.5" × 10" Aquabind MP (Mn-AA) POU: 4.5" × 10" Aquabind EP (AA) POE: 12" column GFH POE: 12" column Alcan FS-50 (Fe-AA)
Unity, Maine	POU RO: Osmonics AG membrane POU: 4.5" × 10" Aquabind MP (Mn-AA)
Carson City, Nev.	POU: 4.5" × 10" GFH POU: 4.5" × 20" Aquabind MP (Mn-AA)
Houston, Texas	POE: 8" column GFH POE: 8" column Alcan FS-50 (Fe-AA)

FIELD SCREENING OF POU AND POE TECHNOLOGIES

A field screening evaluation was performed at the central sites to verify POU and POE treatment unit performance. Four types of adsorptive processes (AA, Fe-AA, Mn-AA, and GFH) and two membrane types (DOW Filmtec and Osmonics AG) were evaluated. Screening evaluations were performed at six different test sites: Tucson, Ariz. (Metro Water); Stagecoach, Nev.; Sun City West, Ariz.; Unity, Maine; Carson City, Nev.; and Houston, Texas. Further information on the testing performed and the ground water quality at each site is presented in the sections below that describe the results of the screening evaluations.

A field testing protocol was developed for the 12-week screening test and is included in Appendix D (including analytical methods, sample preservation requirements, sample tagging and chain of custody procedures). The specifications of the POU/POU units are presented in Table 4.2. The process schematics for five of the six test sites are shown in Figures 4.4 to 4.8.

The field screening program consisted of two phases. Phase A lasted two weeks and was designed to condition the POU units and ensure that sufficient volume was treated to ensure representative performance. Phase B lasted 10 weeks and simulated household operation of the POU units. Table 4.3 shows the sampling schedule for the POU units during Phase A. Table 4.4 shows the sampling and operational schedule during Phase B for the POU units. Table 4.5 shows the sampling and operational schedule for the POE units.

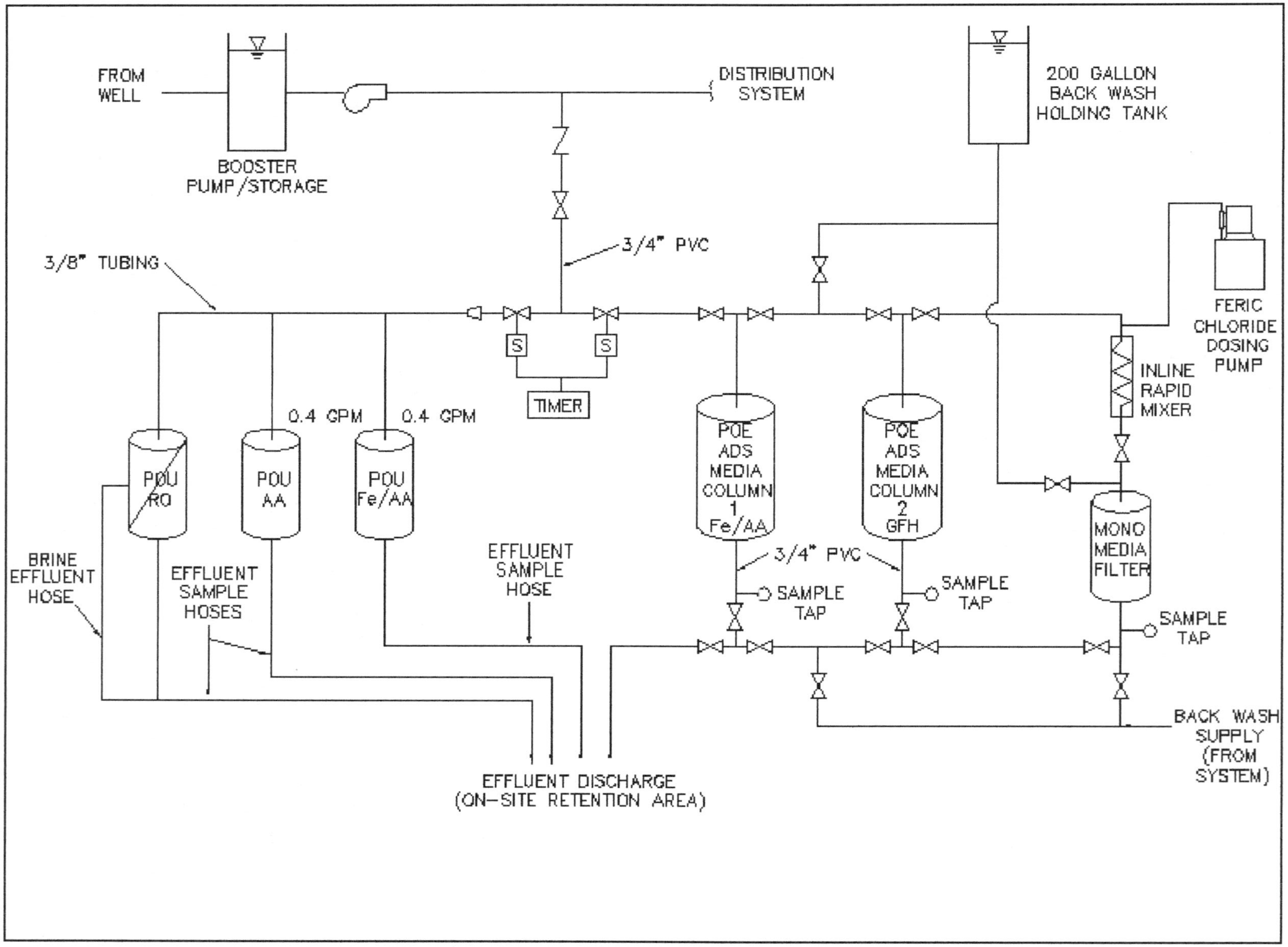

Figure 4.4 Process schematic for Sun City West, Ariz. site

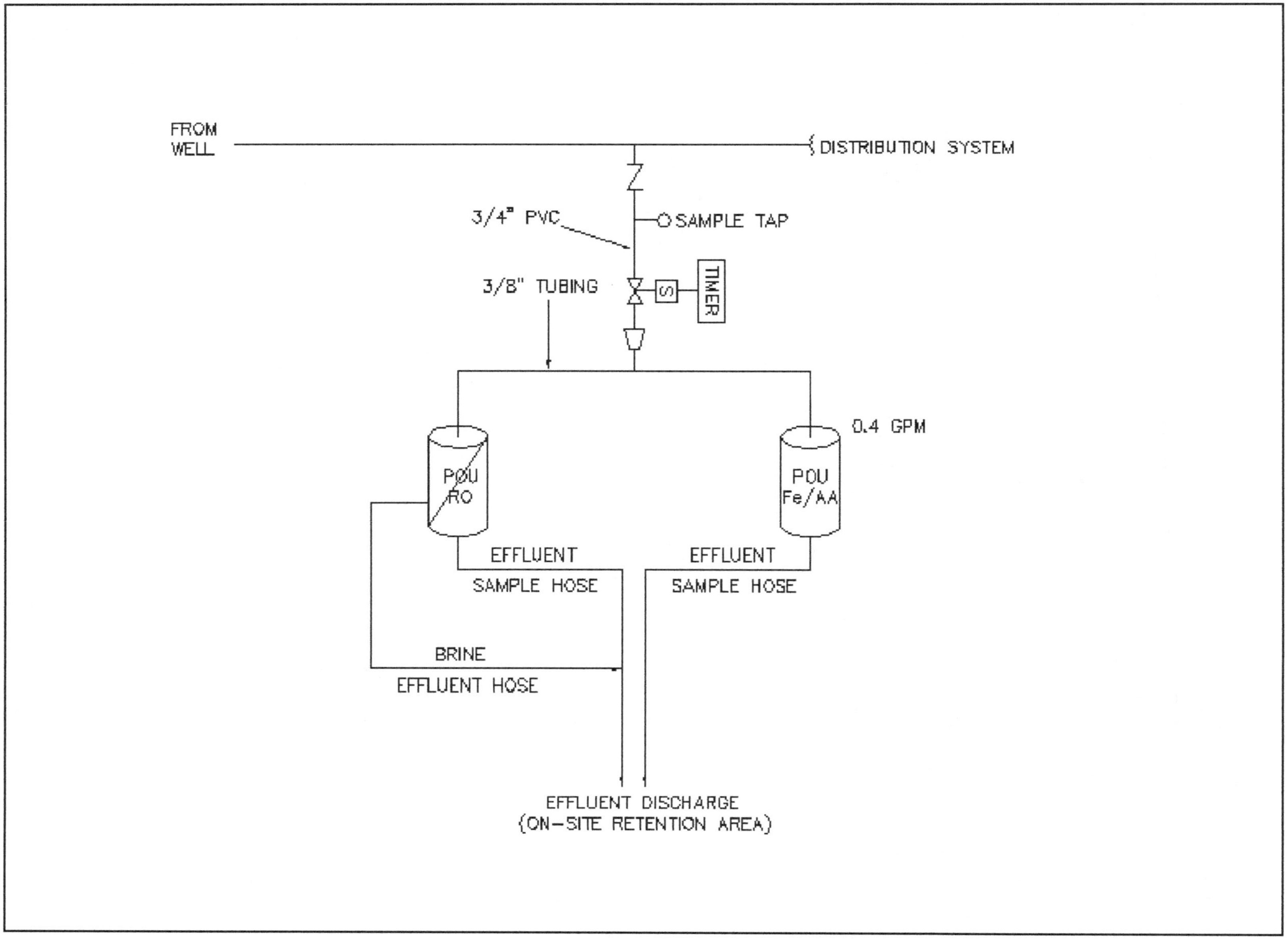

Figure 4.5 Process schematic for Unity, Maine site

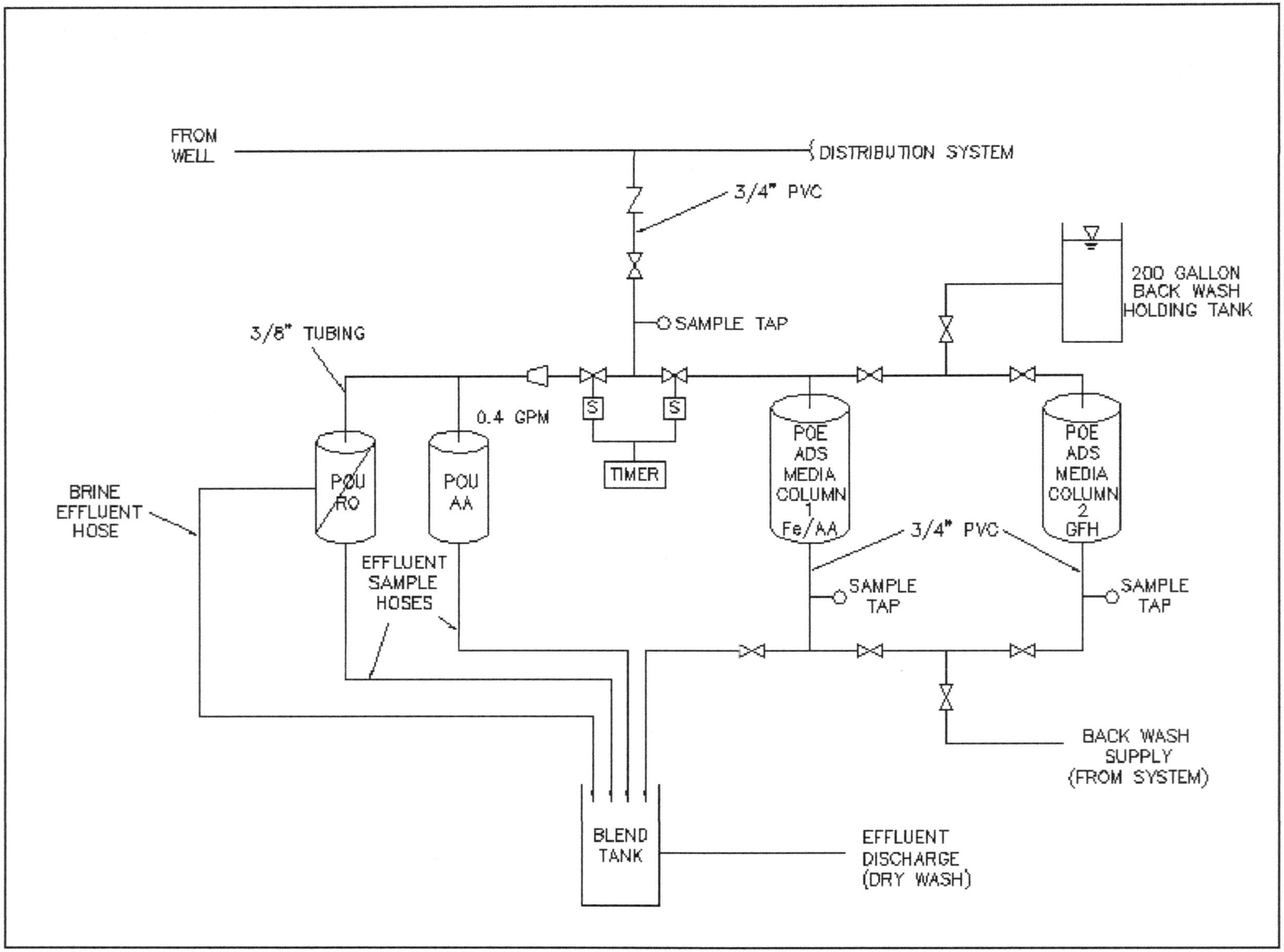

Figure 4.6 Process schematic for Metro Water, Tuscon, Ariz. site

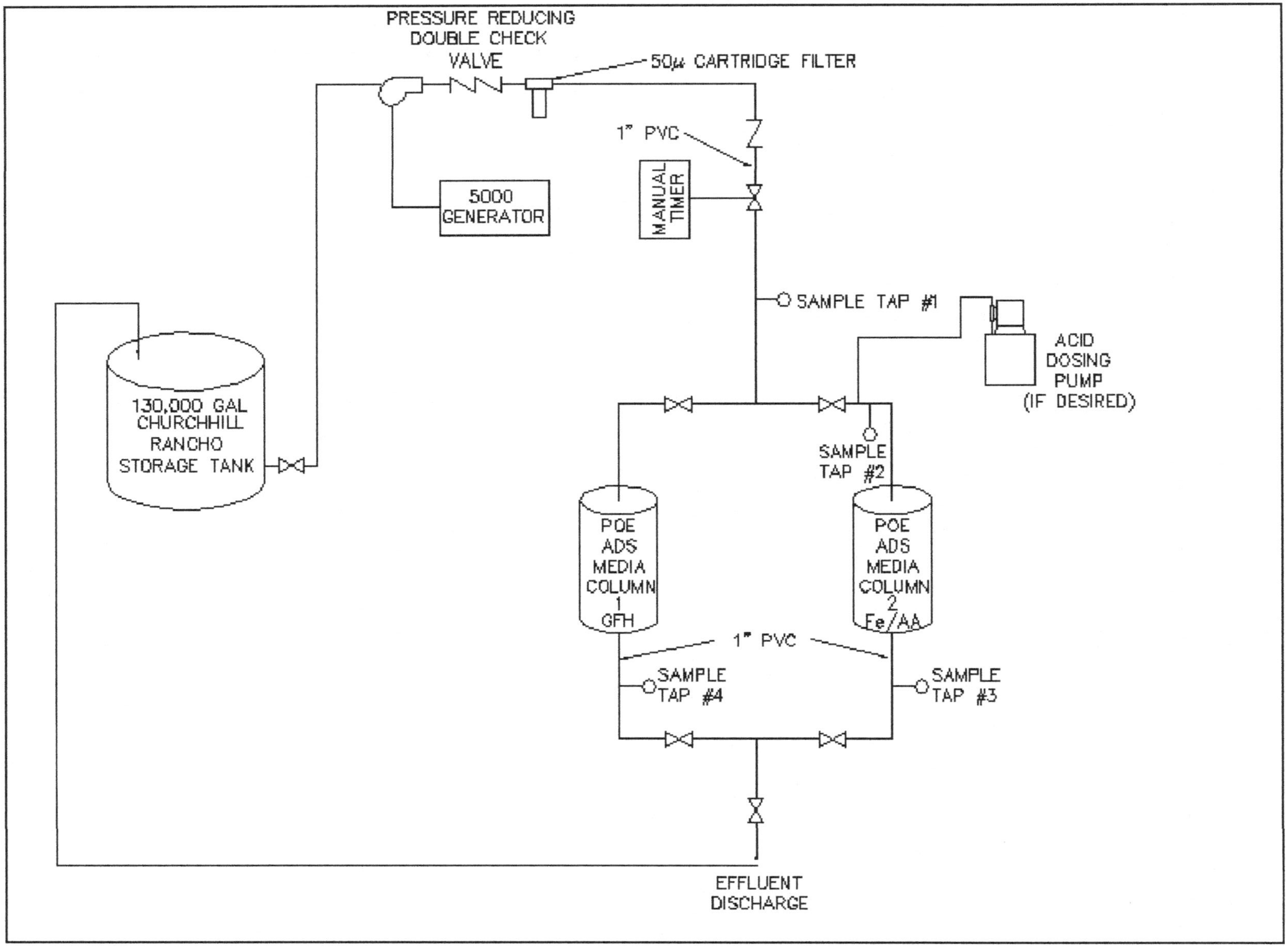

Figure 4.7 Process schematic for Stagecoach, Nev. site

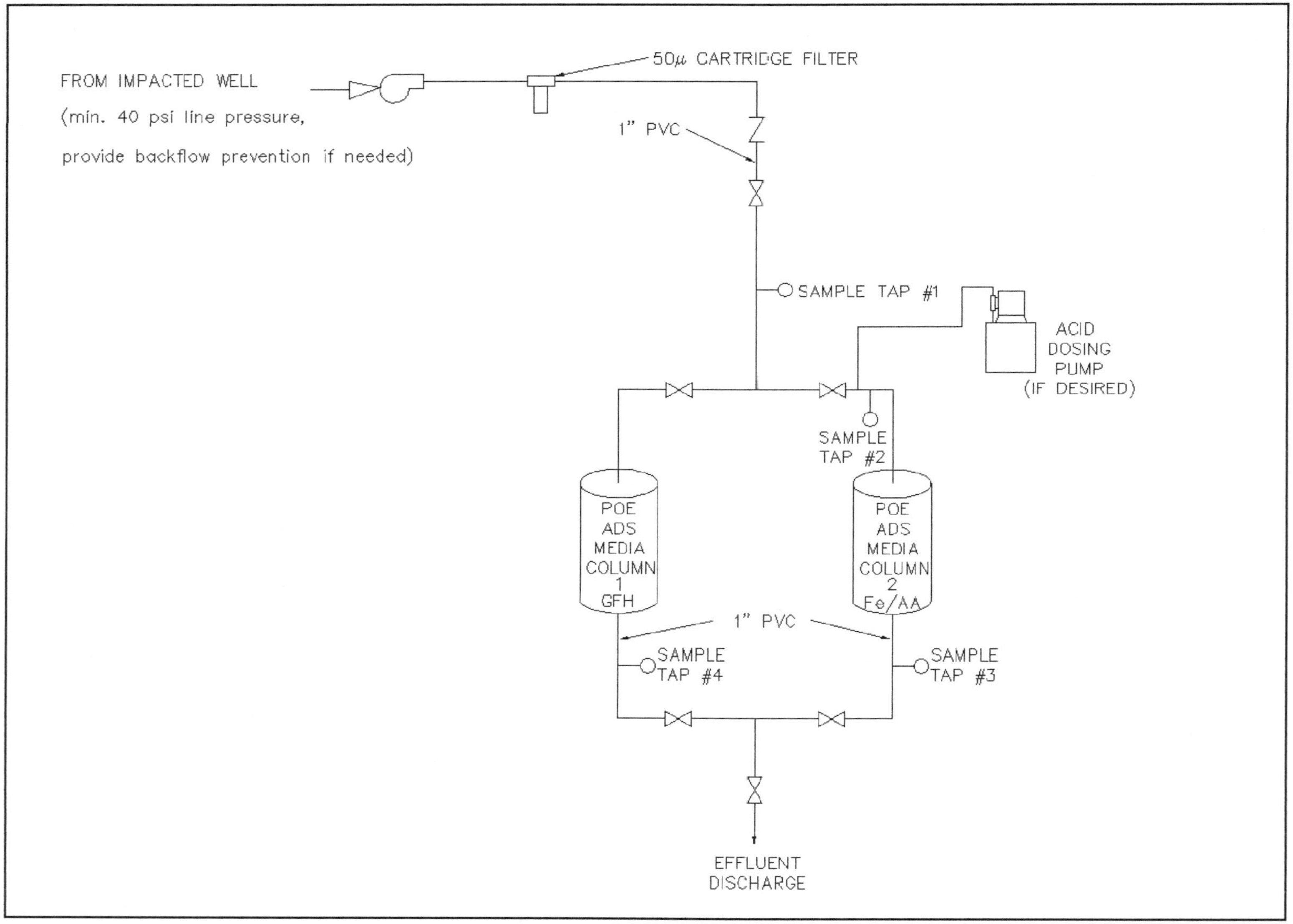

Figure 4.8 Process schematic for Houston, Texas site

Table 4.3
Sampling schedule for POU units during Phase A

Period	Operating cycle	Duration per day	Sample times	Raw water	Treated effluent
First 1-2 weeks	40 min. on/40 min. off	16 hours followed by 8 hour rest under pressure	9:00 AM	Arsenic, TDS, alkalinity, hardness, conductivity, sulfate, nitrate, iron, manganese, fluoride, silica, phosphate, HPC and total coliform (weekly) Temperature, pH, chlorine residual (3 per week)	Arsenic, pH, chlorine residual, conductivity, and silica (3 days per week)
			12:00 PM	—	Arsenic, pH, chlorine residual, conductivity, and silica

Samples were collected three times a week for parameters such as arsenic, silica, pH, chlorine residual, and conductivity from POU/POE treated effluent and once a week from the POU/POE influent (raw water). For the POU units, additional parameters such as alkalinity, TDS, sulfate, fluoride, nitrate, iron, manganese, and phosphate were sampled once a week.

RESULTS OF FIELD SCREENING EVALUATIONS

Metro Water, Tucson, Ariz.

A POU RO system and a POU AA system were installed at this site. Two 12-inch POE columns, one containing GFH and one containing Fe-AA media, were also installed. A limitation on the amount of treated water allowed to be discharged at this site required each of the POE units to be operated at a flow rate of 3 gpm. This established the EBCT of each of the columns at five minutes. The original protocol was for the GFH to be operated at an EBCT of 2.5 minutes. Additionally, difficulties with the control valves on the POE units for the first three weeks of testing resulted in bypassing of the influent to the effluent pipe without treatment. The testing schedule for the POE units was extended to compensate for this equipment malfunction.

Groundwater Quality

Summarized in Table 4.6 is the raw water quality for groundwater from Hub #3 of Metro Water. The raw water was sampled once a week over the test period of 12/10/01 to 2/04/02. The data summarized in Table 4.6 are the averages of the values measured over that period. The average arsenic concentration in the raw water was 0.011 mg/L. The source water had no iron, manganese, or phosphate. The average pH of the raw water was 7.8.

Table 4.4
Sampling and operational schedule for POU units during Phase B

Period	Operating times	Duration (min)	Sampling (weekly)
1st period (2 weeks), Weeks 1-2	6:00 am	2	200 mL first flush (total coliform and HPC) sample
	8:30 am	5	
	11:30 am	5	1 L (Arsenic, TDS, alkalinity, hardness, conductivity, sulfate, nitrate, iron, manganese, fluoride, silica, phosphate, temp., pH, disinfectant residual
	5:30 pm	2	
	6:30 pm	5	
	9:30 pm	3	
2nd period (1 week), Week 3	no flow		1 L before shut down (Arsenic, TDS, alkalinity, hardness, conductivity, sulfate, nitrate, iron, manganese, fluoride, silica, phosphate, temperature, pH, disinfectant residual), 200 mL after startup (total coliform and HPC)
3rd period (7 weeks), Weeks 4-10	6:00 am	2	200 mL first flush (total coliform and HPC)
	8:30 am	5	
	11:30 am	5	1 L (Arsenic, TDS, alkalinity, hardness, conductivity, sulfate, nitrate, iron, manganese, fluoride, silica, phosphate, temperature, pH, disinfectant residual)
	5:30 pm	2	
	6:30 pm	5	

Table 4.5
Sampling and operational schedule for the POE units

Technology	Operational schedule	Raw	Treated effluent
POE Columns: Fe-AA and GFH	Phase A: 16 hours on, 8 hours off	Parameters and frequency same as raw samples for POU units (one set per test site) See Table D3.	Arsenic, pH, chlorine residual, conductivity, and silica (3 days per week)
	Phase B: continuous		

Table 4.6
Source water quality summary from Hub #3 of Tucson, Ariz. water

Parameter	Average	Standard deviation	Maximum value measured	Minimum value measured	Number of samples analyzed
pH	7.8	0.2	8.1	7.4	22
Chlorine residual, mg/L	0.8	0.4	2	0.5	23
Arsenic, mg/L	0.011	0.001	0.013	0.008	15
Alkalinity, mg/L	132.5	7.1	140.0	120.0	8
Fluoride, mg/L	1.5	0.2	1.8	1.1	8
Hardness as $CaCO_3$, mg/L	116.6	15.0	140.0	92.0	9
Iron, mg/L	<0.05	0	<0.05	<0.05	9
Manganese, mg/L	<0.01	0	<0.01	<0.01	9
Nitrate, mg/L	2.1	0.2	2.3	1.7	8
Orthophosphate, mg/L	<0.1	0	<0.1	0.1	8
Sulfate, mg/L	37.6	5.2	43.0	28.0	11
Silica (as SiO_2), mg/L	42.7	0.6	43.0	41.0	15
TDS, mg/L	305.7	35.5	350.0	260.0	7
HPCs, CFUs/mL	31	396	1,160	1.0	8

POU RO Results

Shown in Figure 4.9 are the arsenic and TDS levels in the RO effluent. After treating 780 gallons of water, the POU RO treatment unit effluent showed no detectable levels (<0.002 mg/L) of arsenic (the method detection limit for arsenic was 0.001 mg/L). The POU RO treatment also reduced the silica levels from 42.7 mg/L to <10 mg/L. Approximately 80-90% TDS removal was also observed. The TDS removal efficiency improved with operational time. The data for other water quality parameters for the RO effluent are summarized in Table 4.7. The average HPC level of the RO treated water (135 CFU/mL) was greater than in the feed (source) water (31 CFU/mL). It should be noted that the residual chlorine in the RO feed water was removed by initially passing the water through a granular activated carbon filter.

POU AA Results

The arsenic and silica concentrations in the effluent from the POU AA unit are plotted as functions of bed volumes in Figure 4.10. The POU AA system consistently removed arsenic to non-detectable (<0.002 mg/L) levels until reaching 1,960 bed volumes (2,660 gallons treated). The Aquabind EP media used in the POU AA unit also removed silica from 42.7 mg/L to <10 mg/L. A summary of the water quality of the POU AA treated water is shown in Table 4.8.

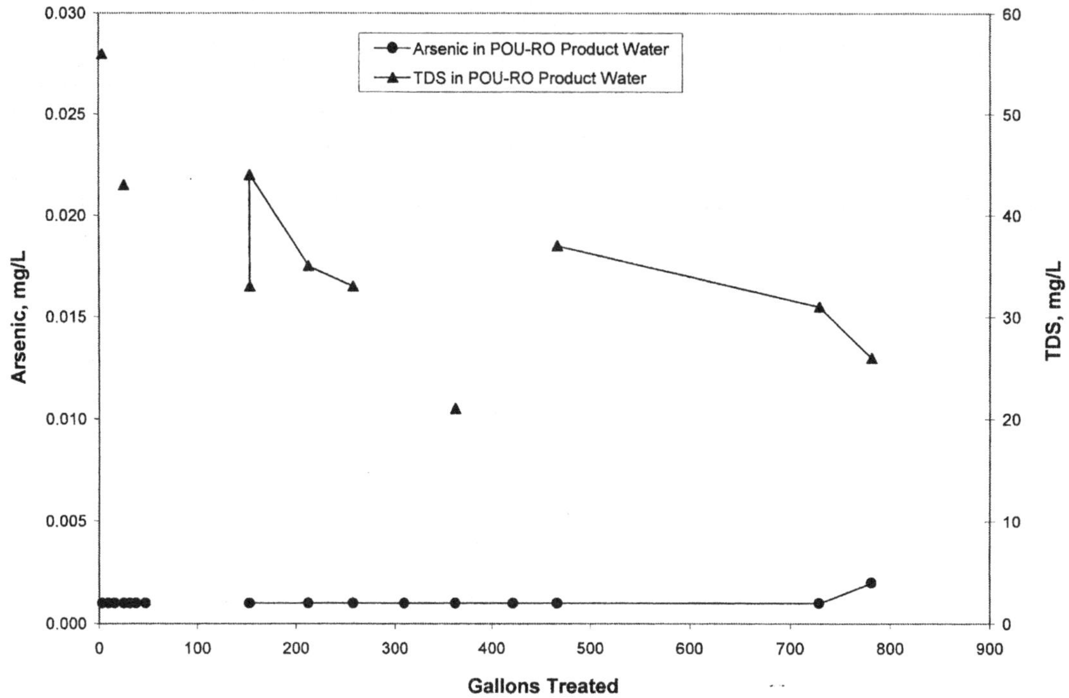

Figure 4.9 Arsenic and TDS levels in Tucson, Ariz. POU RO effluent

Table 4.7
Water quality summary of Tucson, Ariz. POU RO effluent

Parameter	Average	Standard deviation	Maximum value measured	Minimum value measured	Number of samples analyzed
pH	7.6	0.7	8.8	6.7	26
Arsenic, mg/L	<0.002	0	<0.002	<0.001	22
Alkalinity, mg/L	13.5	2.1	18.0	12.0	11
Fluoride, mg/L	<0.4	0	<0.4	<0.4	10
Hardness as $CaCO_3$, mg/L	<13	0	<13	<13	13
Iron, mg/L	<0.05	0	<0.05	<0.05	6
Manganese, mg/L	<0.01	0	<0.01	<0.01	7
Nitrate, mg/L	<0.2	0	<0.2	<0.2	5
Orthophosphate, mg/L	<0.1	0	<0.1	<0.1	5
Sulfate, mg/L	<2.0	0	<2.0	<2.0	10
Silica (as SiO_2), mg/L	6.5	0.7	7.7	4.8	22
TDS, mg/L	35.9	9.9	56.0	21.0	10
HPCs, CFUs/mL	135	85	260	1	8

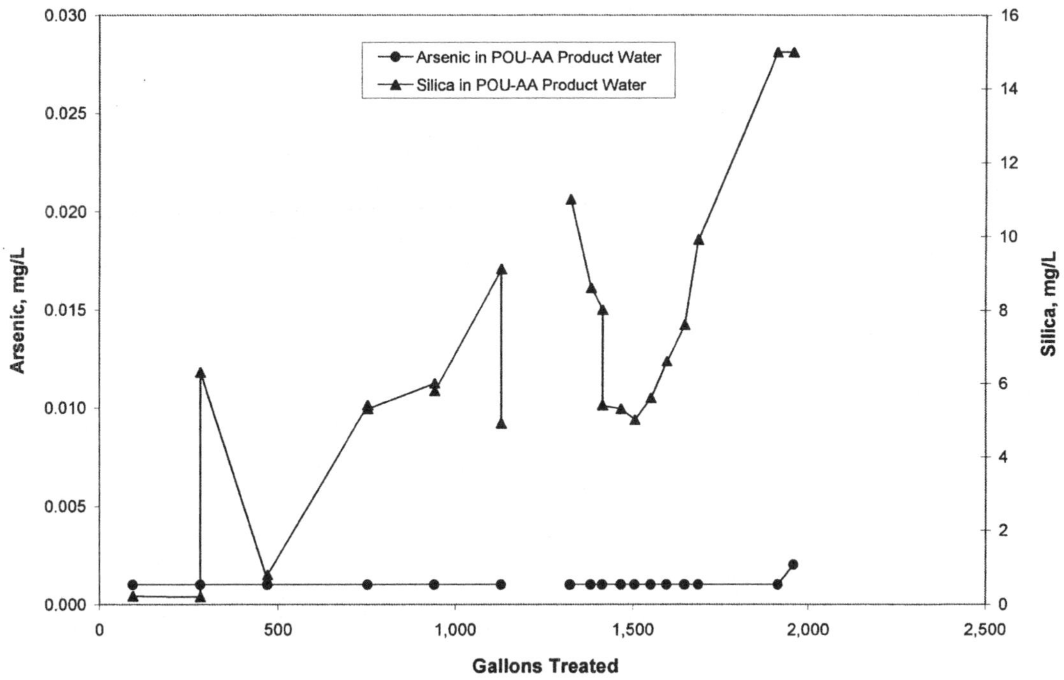

Figure 4.10 Arsenic and silica levels in Tucson, Ariz. POU AA effluent

Table 4.8
Water quality summary of Tucson, Ariz. POU AA effluent

Parameter	Average	Standard deviation	Maximum value measured	Minimum value measured	Number of samples analyzed	BVs to breakthrough (if applicable)
pH	7.8	0.3	8.6	7.4	26	
Arsenic, mg/L	<0.001	0	<0.002	<0.001	23	2,660 (non-detect levels)
Alkalinity, mg/L	138.3	7.2	150.0	120.0	12	
Fluoride, mg/L	0.49	0.13	0.79	0.40	11	>2,660
Hardness as $CaCO_3$, mg/L	120	10.0	130.0	110.0	9	
Iron, mg/L	<0.05	0	<0.06	<0.05	7	
Manganese, mg/L	<0.01	0	<0.01	<0.01	7	
Nitrate, mg/L	2.2	0.2	2.4	2.0	6	
Orthophosphate, mg/L	<0.1	0	<0.1	<0.1	7	
Sulfate, mg/L	38.5	2.2	41.0	35.0	6	
Silica (as SiO_2), mg/L	6.4	3.9	15.0	0.2	23	>2,660
TDS, mg/L	273.3	15.1	290.0	250.0	6	
HPCs, CFUs/mL	1.4	1.0	4.0	1.0	9	

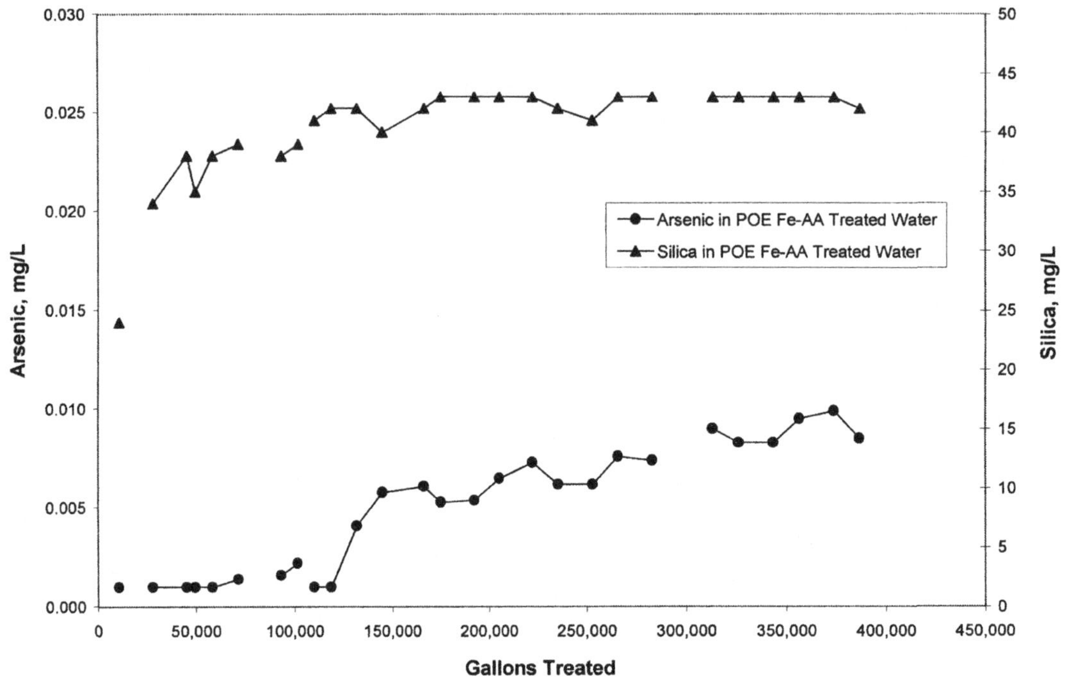

Figure 4.11 Arsenic and silica levels in Tucson, Ariz. POE Fe-AA effluent

Some removal of fluoride by POU AA treatment was also observed. As expected, no TDS removal was observed in the POU AA effluent. The average HPC level in the POU AA treated water (1.3 CFUs/mL) was lower than the average levels in both the source water (31 CFUs/mL) and in the POU RO treated water (40 CFUs/mL). The varying operational conditions (simulating intermittent intensive use and vacation periods) did not appear to affect the POU AA performance.

POE Fe-AA and GFH Results

Breakthrough curves for arsenic and silica levels in the effluent from the POE Fe-AA and POE-GFH units are shown in Figures 4.11 and 4.12, respectively. Some decrease was observed in silica levels. This decrease is more dominant in the initial periods, especially for the Fe-AA column. Summaries of POE Fe-AA and POE GFH treated water quality are shown in Tables 4.9 and 4.10, respectively. No significant removal of sulfate and fluoride was observed in the POE Fe-AA unit.

Figures 4.13 and 4.14 show the arsenic concentrations in the raw and treated waters for the POE Fe-AA and GFH columns, respectively. As shown in Figure 4.13, the breakthrough (arsenic level less than 0.010 mg/L) for Fe-AA occurred at approximately 23,760 BVs (356,400 gallons treated). The GFH system removed arsenic to 0.006 mg/L until 45,800 BVs (343,400) as shown in Figure 4.14.

Sun City West, Ariz. (Phoenix Metropolitan Area)

One POU RO unit and one POU unit with Mn-AA media were installed at a Sun City West, Ariz. well site (Well 28B) and each was operated using an automatic timer. The Phase A

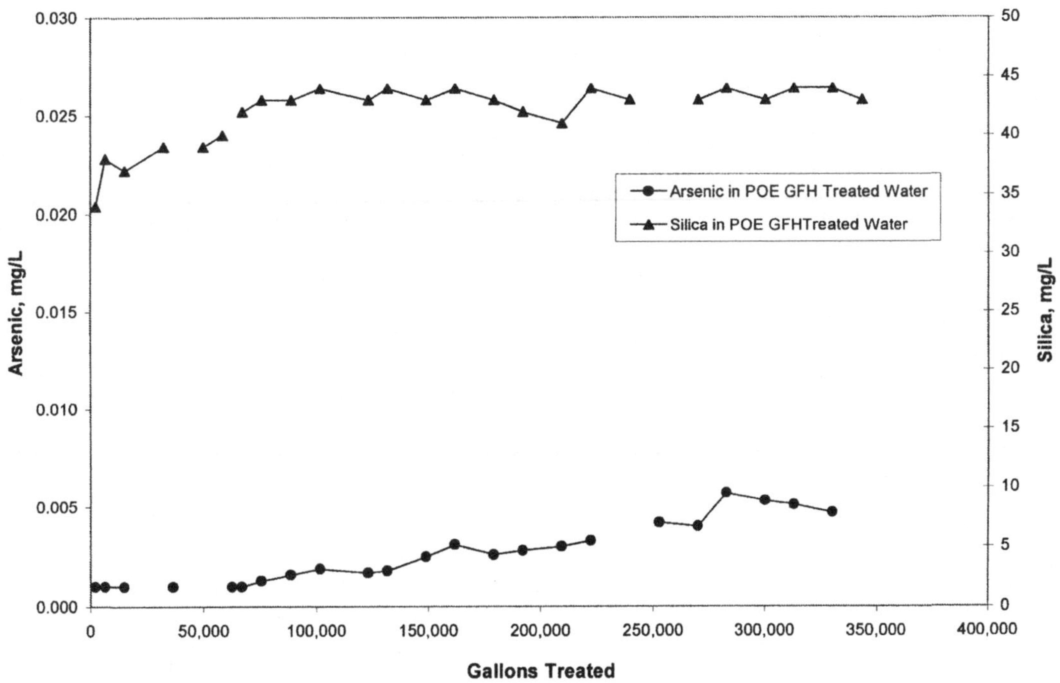

Figure 4.12 Arsenic and silica levels in Tucson, Ariz. POE GFH effluent

Table 4.9
Water quality summary of Tucson, Ariz. POE Fe-AA effluent

Parameter	Average	Standard deviation	Maximum value measured	Minimum value measured	Number of samples analyzed	BVs to breakthrough (if applicable)
Arsenic, mg/L	0.005	0.003	0.010	<0.001	27	23,760
Aluminum, mg/L	< 0.1	0	<0.1	<0.1	9	
Fluoride, mg/L	1.4	0.1	1.6	1.3	9	NA*
Sulfate, mg/L	42	3	48	36	9	
Silica as SiO_2, mg/L	40	4	43	24	27	11,100

* not available

protocol originally stipulated an operational cycle of 40 minutes on and 40 minutes off for 16 hours followed by 8 hours off while maintaining pressure. The protocol for this site for the first two weeks of testing (Phase A) differed from the original protocol as follows. For the first two weeks of testing, the POU Mn-AA and RO units at this site were run continuously to establish a breakthrough curve for the smaller 10-inch columns and to achieve the maximum number of gallons treated through the POU RO unit during the two weeks prior to Phase B. Following breakthrough of the two columns, the media cartridges were replaced and Phase B was conducted to

Table 4.10
Water quality summary of Tucson, Ariz. POE GFH effluent

Parameter	Average	Standard deviation	Maximum value measured	Minimum value measured	Number of samples analyzed	BVs to breakthrough (if applicable)
Arsenic, mg/L	0.003	0.002	0.006	0.001	24	>45,800
Iron, mg/L	0.05	0.004	0.06	<0.05	9	
Manganese, mg/L	<0.01	0	<0.01	<0.01	2	
Sulfate, mg/L	42	3	47	36	9	
Silica as SiO_2, mg/L	42	2	44	37	24	8,900

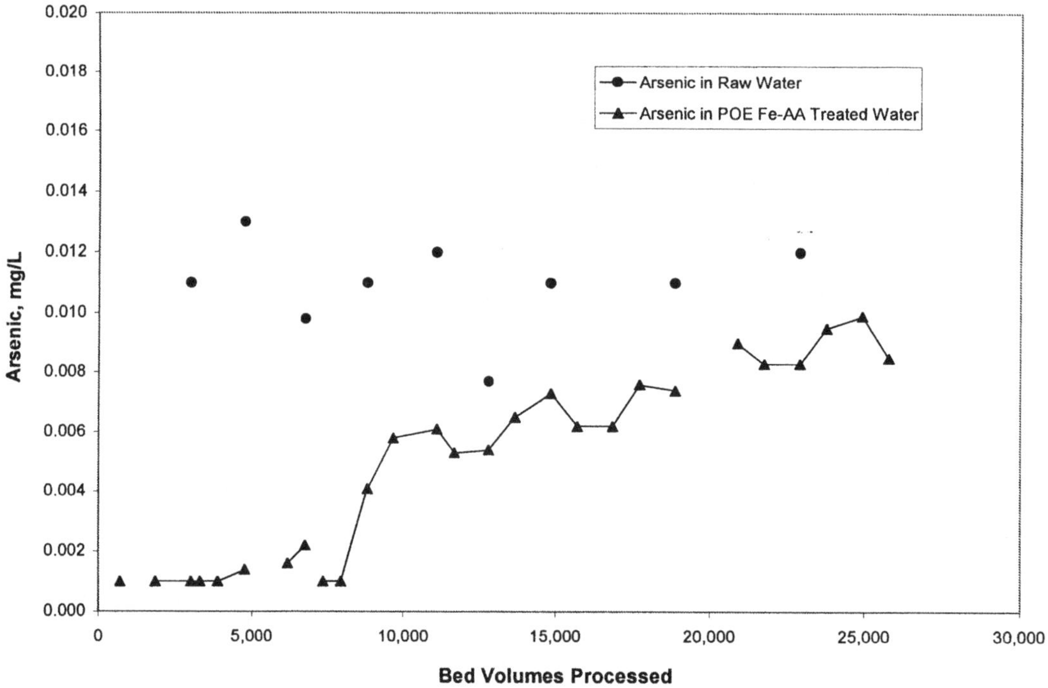

Figure 4.13 Arsenic levels in Tucson, Ariz. raw and POE Fe-AA treated effluent as a function of BVs processed

simulate home use. Two 12-inch POE columns, one with GFH and one with Fe-AA, were also installed at this site.

Groundwater Quality

Table 4.11 summarizes the raw water quality for groundwater from Well 28B at Sun City West. The raw water was sampled once a week over the test period of 1/14/02 to 4/24/02. The data

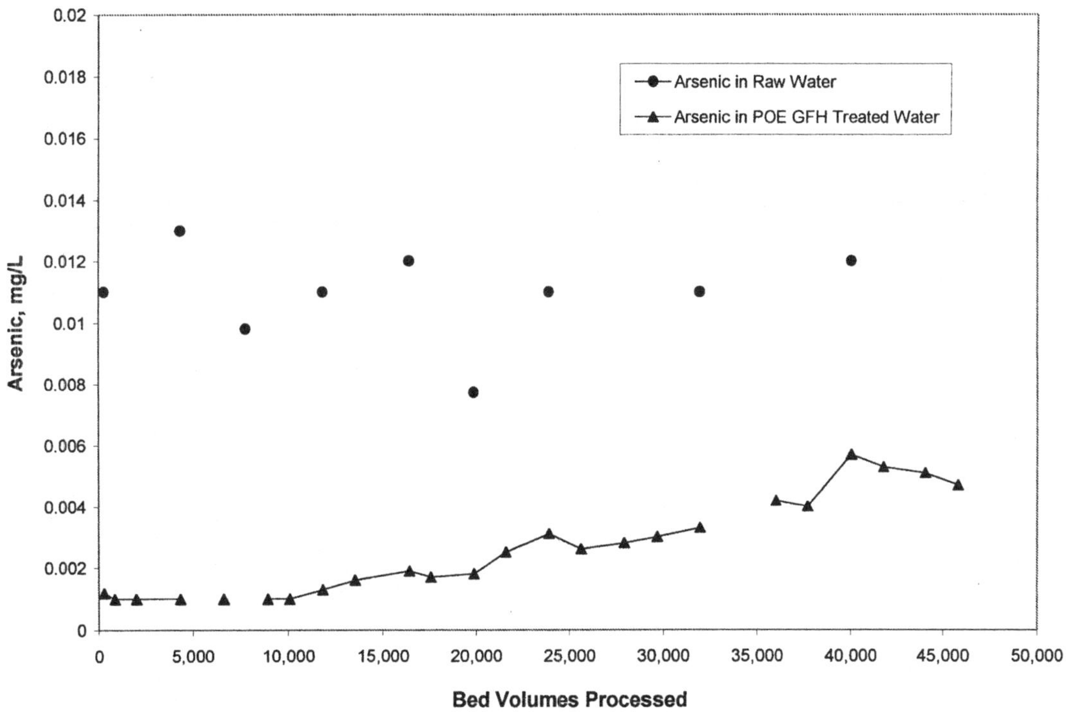

Figure 4.14 Arsenic levels in Tucson, Ariz. raw and POE GFH treated effluent as a function of BVs processed

summarized in Table 4.11 are the averages of the values measured over that period. The average arsenic level in the raw water was 0.023 mg/L. The source water had low levels of iron, manganese, and phosphorus. The average pH of the raw water was high at 8.4.

POU RO Results

Shown in Figure 4.15 are the arsenic, silica, and TDS levels in the RO effluent. Even after treating 1,300 gallons of water, the POU RO effluent showed very low levels (0.003 mg/L) of arsenic. The POU RO treatment also lowered the silica (as SiO_2) levels from 12.9 mg/L to <5 mg/L. Approximately 92-95% TDS removal was also observed. The data for the rest of the water quality parameters for the RO product water are summarized in Table 4.12. The HPC levels in the RO effluent were substantially higher (933 cfu/mL) than in the RO influent (396 cfu/mL)

POU AA Results

The arsenic and silica concentrations in the POU AA unit are shown in Figure 4.16. The POU AA system consistently removed arsenic to non-detectable (<0.001 mg/L) levels until approximately 2,600 BVs (1,780 gallons treated). The Aquabind EP media in the POU AA unit also removed silica to a limited extent, from 26.2 mg/L to <10 mg/L (as SiO_2) prior to arsenic breakthrough. A summary of POU AA treated water is shown in Table 4.13. The HPC data for the POU AA treated water did not show a significant difference from the raw water. The varying operational conditions (simulation of intermittent intensive use and vacation) did not affect the POU AA performance.

Table 4.11
Source water quality summary from Well 28B, Sun City West, Ariz.

Parameter	Average	Standard deviation	Maximum value measured	Minimum value measured	Number of samples analyzed
pH	8.4	0.1	8.6	8.2	13
Chlorine residual, mg/L	0.83	0.13	1.13	0.62	13
Arsenic, mg/L	0.023	0.001	0.023	0.021	13
Alkalinity, mg/L	146.9	2.3	150.0	142.0	13
Fluoride, mg/L	1.8	0.2	2.2	1.2	13
Hardness as $CaCO_3$, mg/L	66.5	6.1	77.0	54.0	13
Iron, mg/L	0.04	0.09	0.32	0.01	12
Manganese, mg/L	0.02	0.02	0.09	0.01	12
Nitrate, mg/L	4.3	0.3	4.9	3.8	13
Total Phosphorus, mg/L	<0.05	0	<0.05	<0.05	12
Sulfate, mg/L	42.8	1.3	45.7	41.5	13
Silica (as SiO_2), mg/L	12.9	0.9	14.2	11.4	13
TDS, mg/L	305.7	14.2	327.0	277.0	13
HPCs, CFUs/mL	396	754	2000	<1	13

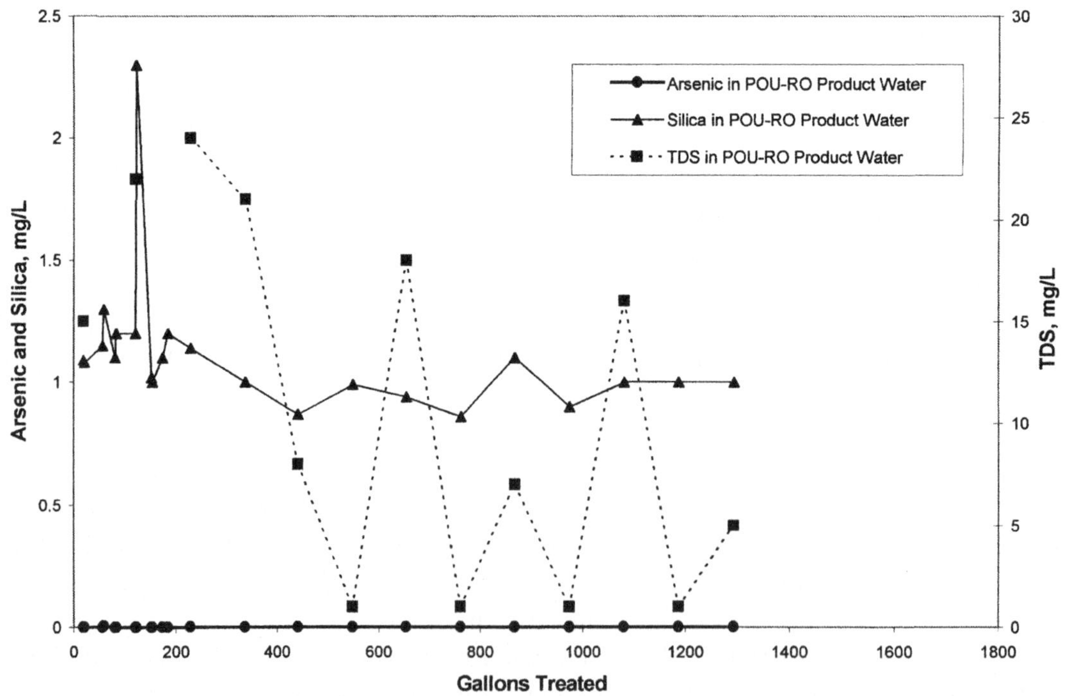

Figure 4.15 Arsenic, silica, and TDS levels in Sun City West, Ariz. POU RO effluent

Table 4.12
Water quality summary of Sun City West, Ariz. POU RO effluent

Parameter	Average	Standard deviation	Maximum value measured	Minimum value measured	Number of samples analyzed
pH	7.8	0.4	8.7	7.1	23
Arsenic, mg/L	0.002	0.001	0.007	0.001	23
Alkalinity, mg/L	10.1	0.3	11.0	10.0	13
Fluoride, mg/L	0.10	0.03	0.20	0.10	11
Hardness as $CaCO_3$, mg/L	<7	0	<7	<7	11
Iron, mg/L	0.006	0.005	0.020	0.005	11
Manganese, mg/L	0.28	0.46	1.0	0.01	11
Nitrate, mg/L	0.57	0.12	0.70	0.30	11
Total Phosphorus, mg/L	<0.15	0	<0.15	<0.15	11
Sulfate, mg/L	<5	0	<5	<5	13
Silica (as SiO_2), mg/L	1.1	0.3	2.3	0.9	23
TDS, mg/L	10.8	8.9	24.0	1.0	13
HPCs, CFUs/mL	933	958	>2,000	<1	12

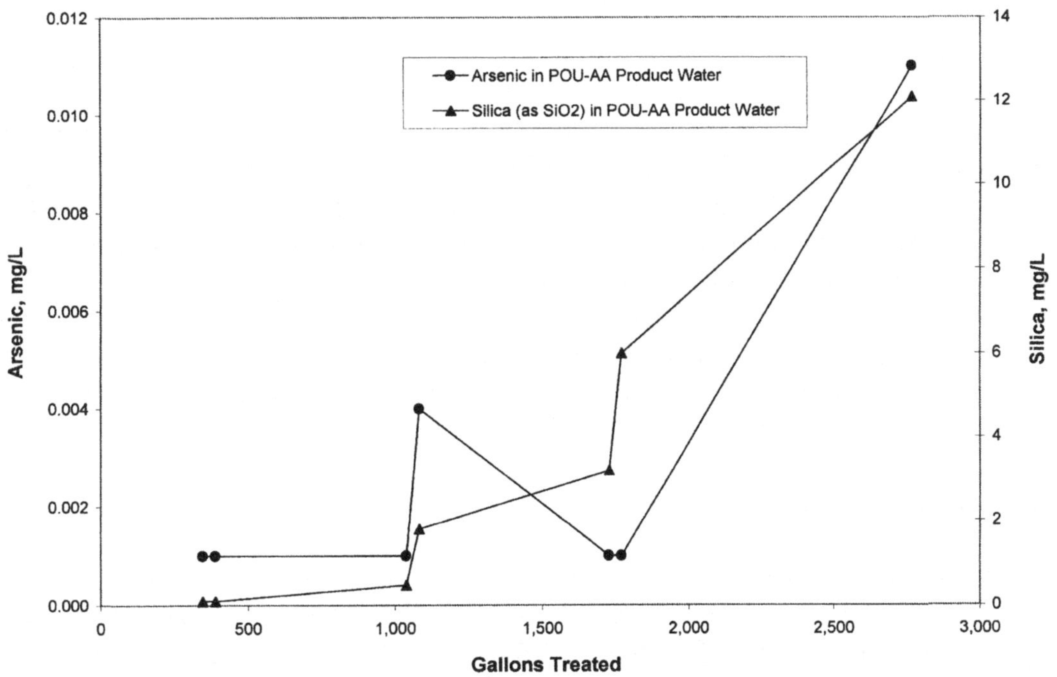

Figure 4.16 Arsenic and silica levels in Sun City West, Ariz. POU AA effluent

Table 4.13
Water quality summary of Sun City West, Ariz. POU AA effluent

Parameter	Average	Standard deviation	Maximum value measured	Minimum value measured	Number of samples analyzed	BVs to breakthrough (if applicable)
pH	8.1	0.2	8.4	7.7	23	
Arsenic, mg/L	0.012	0.009	0.025	<0.001	25	2,600
Alkalinity, mg/L	149.3	4.1	156.0	140.0	13	
Fluoride, mg/L	1.5	0.6	1.9	0.1	11	None*
Hardness as $CaCO_3$, mg/L	66.4	5.9	75.0	54.0	11	
Iron, mg/L	0.014	0.005	0.020	<0.010	11	
Manganese, mg/L	<0.011	<0.003	<0.020	<0.010	11	
Nitrate, mg/L	4.1	0.2	4.4	3.7	11	
Total Phosphorus, mg/L	0.15	0.01	0.18	0.15	11	
Sulfate, mg/L	43.0	1.7	46.4	41.3	11	
Silica (as SiO_2), mg/L	8.8	5.4	14.9	<0.1	23	5,100
TDS, mg/L	280	46	320	135	13	
HPCs, CFUs/mL	201	569	>2,000	<1	12	

* no fluoride removal was observed

POU Mn-AA Results

The arsenic and silica concentrations from testing of the POU Mn-AA unit shown in Figure 4.17. The POU Mn-AA unit consistently removed arsenic to non-detectable (<0.002 mg/L) levels until reaching 2,600 BVs (1,780 gallons treated). As shown in Figure 4.17, the POU Mn-AA unit showed a very sharp breakthrough after initial leakage. The Aquabind MP media used in the POU Mn-AA unit also removed silica (as SiO_2) to a limited extent, from 12.9 mg/L to less than 10 mg/L prior to arsenic breakthrough. A summary of the water quality of the POU Mn-AA effluent is shown in Table 4.14. The HPC levels in the POU Mn-AA treated water showed no significant difference from the levels in the raw water. The varying operational conditions (simulating intermittent intensive use and vacation periods) did not affect the POU Mn-AA performance.

POE Fe-AA and GFH Results

Breakthrough curves for arsenic levels in POE Fe-AA and POE GFH treated water are shown in Figures 4.18 and 4.19, respectively. Summaries of POE Fe-AA and POE GFH treated water are shown in Tables 4.15 and 4.16, respectively. The POE Fe-AA unit removed more silica than the GFH unit. The Fe-AA column removed arsenic below 10 ppb breakthrough until 4,200 BVs (63,400 gallons treated). The GFH column removed arsenic below 10 ppb breakthrough until 24,500 BVs (368,600 gallons treated).

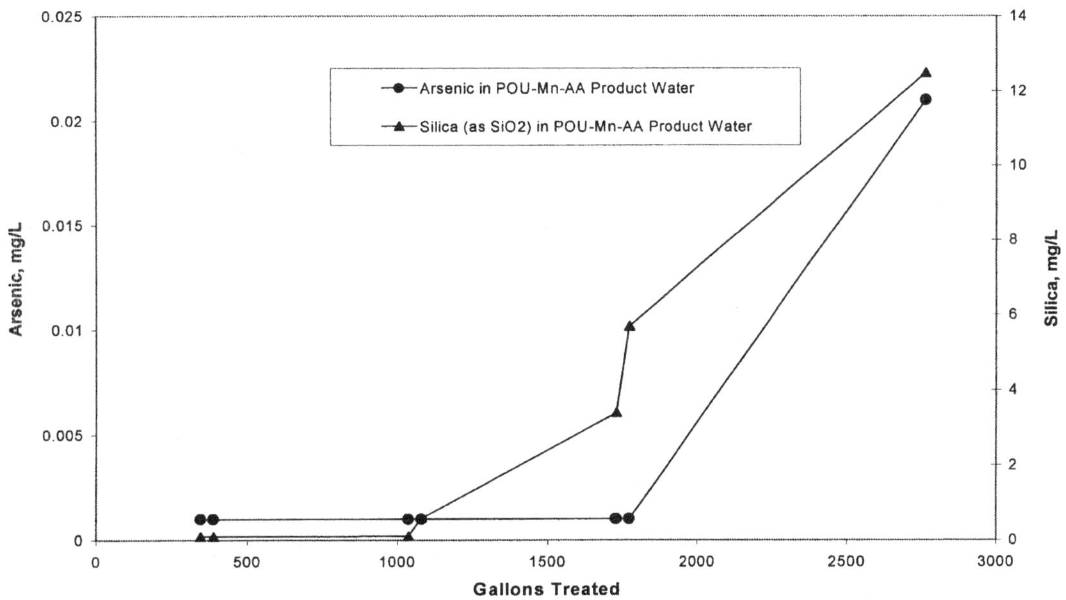

Figure 4.17 Arsenic and silica levels in Sun City West, Ariz. POU Mn-AA effluent

Table 4.14
Water quality summary of Sun City West, Ariz. POU Mn-AA effluent

Parameter	Average	Standard deviation	Maximum value measured	Minimum value measured	Number of samples analyzed	BVs to breakthrough (if applicable)
pH	8.2	0.1	8.5	7.9	23	
Arsenic, mg/L	0.014	0.011	0.026	<0.001	25	2,600
Alkalinity, mg/L	146	6	153	129	13	
Fluoride, mg/L	1.5	0.6	2.0	0.1	13	None*
Hardness as $CaCO_3$, mg/L	66	10	89	47	11	
Iron, mg/L	0.01	0.01	0.02	0.01	11	
Manganese, mg/L	0.01	0.01	0.03	<0.01	11	
Nitrate, mg/L	4.1	0.2	4.4	3.7	11	
Total Phosphorus, mg/L	0.15	0.01	0.18	<0.15	11	
Sulfate, mg/L	41.7	5.0	46.7	25.9	13	
Silica (as SiO_2), mg/L	8.6	5.6	14.2	<0.1	23	5,100
TDS, mg/L	287	32	335	217	12	
HPCs, CFUs/mL	205	567	>2,000	<1	12	

* no fluoride removal was observed

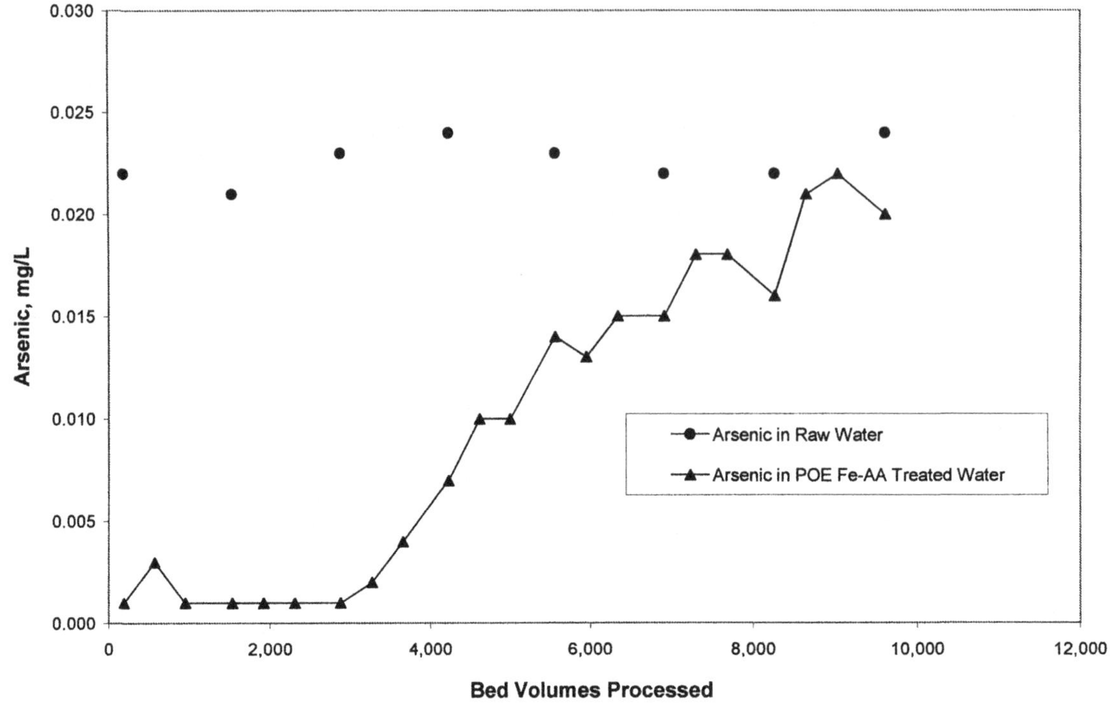

Figure 4.18 Arsenic levels in Sun City West, Ariz. raw and POE Fe-AA treated effluent as a function of BVs processed

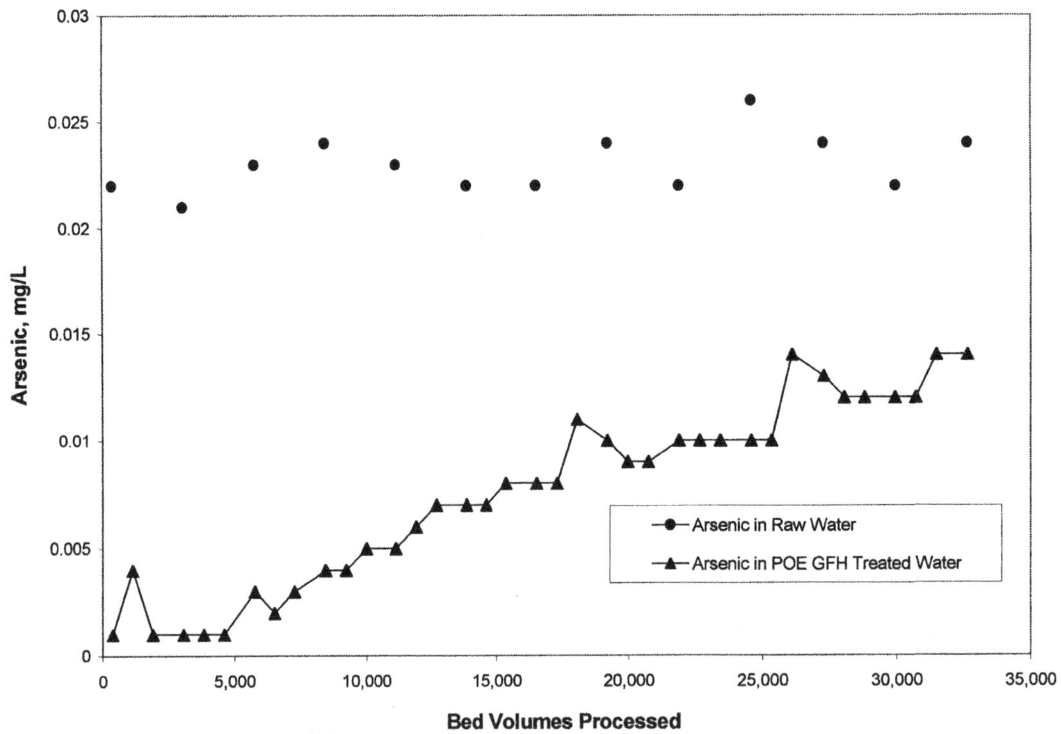

Figure 4.19 Arsenic levels in Sun City West, Ariz. raw and POE GFH treated effluent as a function of BVs processed

Table 4.15
Water quality summary of Sun City West, Ariz. POE Fe-AA effluent

Parameter	Average	Standard deviation	Maximum value measured	Minimum value measured	Number of samples analyzed	BVs to breakthrough (if applicable)
pH	8.2	0.3	8.5	7.2	21	
Arsenic, mg/L	0.010	0.008	0.022	<0.001	22	4,200
Silica, mg/L	9.6	4.0	13.6	1.3	22	5,600

Table 4.16
Water quality summary of Sun City West, Ariz. POE GFH effluent

Parameter	Average	Standard deviation	Maximum value measured	Minimum value measured	Number of samples analyzed	BVs to breakthrough (if applicable)
pH	8.3	0.3	8.5	7.2	36	
Arsenic, mg/L	0.008	0.004	0.014	<0.001	37	24,500
Silica, mg/L	12.0	3.3	15.4	0.1	37	9,200

Stagecoach, Nev.

Two POE units were installed in October 2001 at the Stagecoach, Nev. screening site, at Well 2 of the Churchill Ranchos Water System. The media tested in the POE columns were GFH (supplied by US Filter) and Fe-AA (trade name FS-50 manufactured by Alcan). Training on sampling frequency, locations, field measurements, and troubleshooting was provided to Stagecoach General Improvement District personnel by the research team and a local representative of the POE vendor. Due to difficulties encountered with disposal of the treated water at the Well 2 site of the Churchill Ranchos Water System, the POE units were relocated to Stagecoach's 130,000 gallon storage tank. Water loss and minimizing treated water discharge were concerns at this site. As a result, the effluent from the column with the Fe-AA media, which is NSF approved, was returned to the storage tank. Water was pumped from the storage tank through an approved reduced pressure double check backflow prevention device, treated through the Fe-AA column, and then returned to the storage tank. Since the GFH media did not have NSF approval during the testing phase of this project, the GFH treated water was transferred and disposed to a nearby wash. Testing was initiated in March 2002 at the Stagecoach, Nev. site. Laboratory analyses for the 12-week screening study were performed by the Nevada State Laboratory.

Groundwater Quality

The raw water quality for the Stagecoach site is summarized in Table 4.17. The raw water was sampled for the first week of testing beginning 1/24/02 prior to shutdown of the system. The

Table 4.17
Source water quality summary from Stagecoach, Nev.

Parameter	Result (average values)	Number of samples analyzed
pH	8.2	2
Conductivity, µS	315	2
Arsenic, mg/L	0.024 (maximum)	3
Alkalinity, mg/L	110	3
Fluoride, mg/L	0.13	3
Hardness as $CaCO_3$, mg/L	112.5	3
Iron, mg/L	0.73	3
Manganese, mg/L	0.006	1
Nitrate as N, mg/L	0.35	2
Total Phosphorus, mg/L	0.020	2
Sulfate, mg/L	50	3
Silica (as SiO_2), mg/L	18.5	2
TDS, mg/L	247	3
HPCs, CFUs/mL	228	1

arsenic level in the raw water was 0.024 mg/L. The raw water had low levels of manganese and total phosphorus and varying levels of iron, from 0.05 to 2.1 mg/L. The seasonally varying levels of iron in the groundwater gradually plugged the adsorption units, since the iron is oxidized. A 50-micron prefilter was installed at this site prior to the POE units to help prevent clogging of the adsorption columns. The average pH of the raw water was 8.2.

POE Fe-AA and GFH Results

Breakthrough curves for the Fe-AA and GFH units are shown in Figures 4.20 and 4.21, respectively. The POE Fe-AA unit was operated with an EBCT of 5 minutes for 16 hours, followed by 8 hours of rest. The flow through the contactor was 3 gpm. The Fe-AA treated approximately 2,300 BVs (34,600 gallons) to arsenic breakthrough. The GFH column was operated with a 2.5 minute EBCT for the same duration as the Fe-AA unit. The GFH treated 7,300 BVs or 110,000 gallons before the adsorption media was plugged with iron. The vessel was backwashed and the adsorption capacity was recovered. After backwashing, the GFH vessel treated approximately 4,500 additional BVs prior to breakthrough. Tables 4.18, 4.19, and 4.20 summarize the effluent sampling data for each of the three tests. Based on the breakthrough curves, both of these media were determined to be suitable for POU application. For POE application to be successful, the vessels would require periodic backwashing.

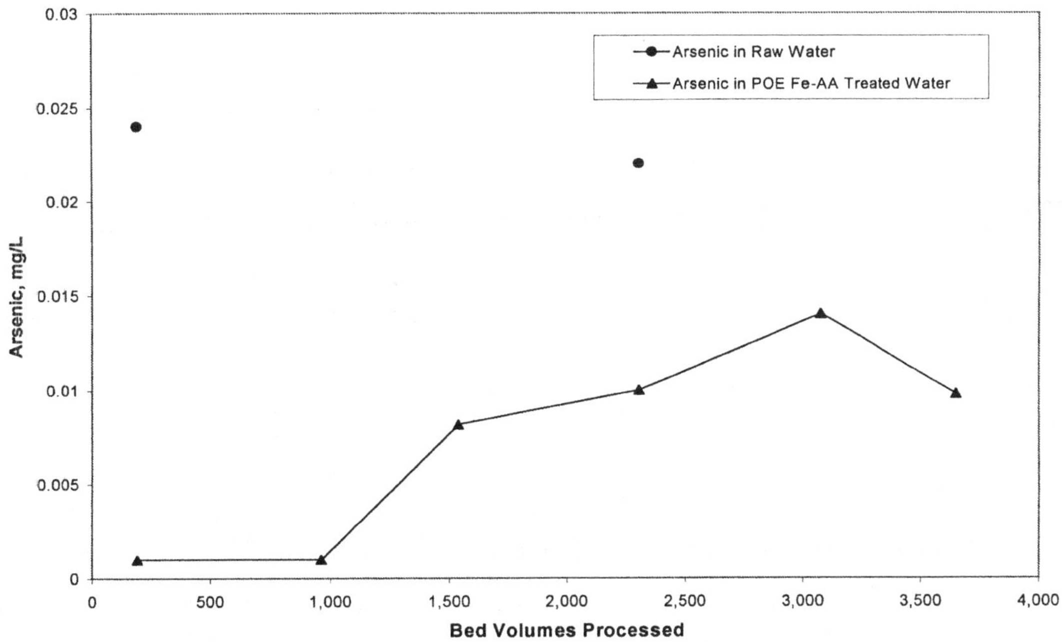

Figure 4.20 Arsenic levels in Stagecoach, Nev. raw and POE Fe-AA treated effluent as a function of BVs processed

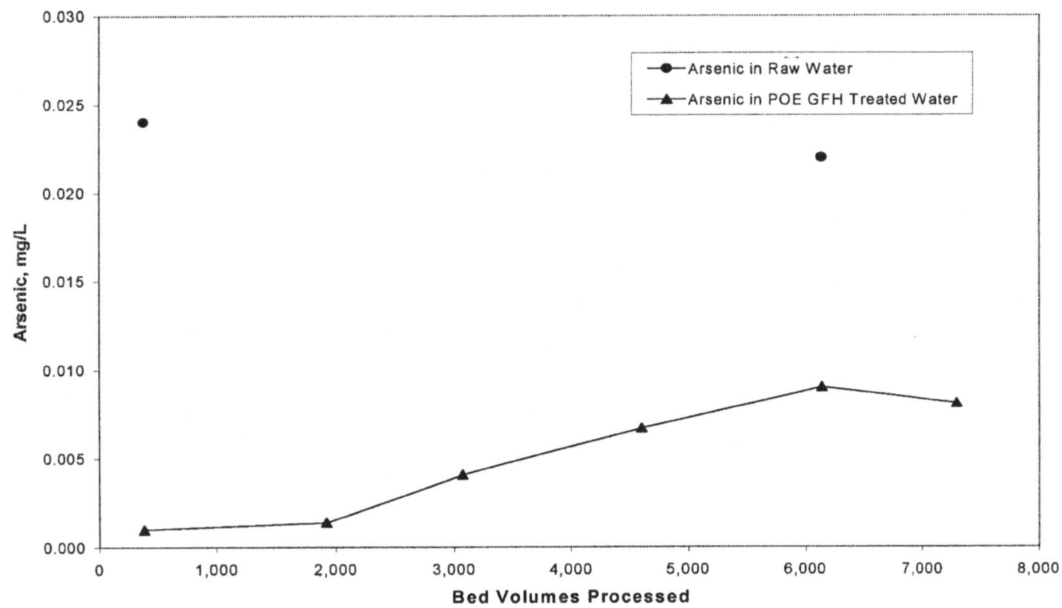

Figure 4.21 Arsenic levels in Stagecoach, Nev. raw and POE GFH treated effluent as a function of BVs processed

Table 4.18
Water quality summary of Stagecoach, Nev. POE Fe-AA effluent

Parameter	Average	Standard deviation	Maximum value measured	Minimum value measured	Number of samples analyzed	BVs to breakthrough (if applicable)
Arsenic, mg/L	0.007	0.005	0.014	<0.001	6	2,300
Iron, mg/L	<0.05	0	<0.05	<0.05	2	
Silica, mg/L	19.9	8.7	26.0	1.2	6	NA*

* not applicable (as raw water silica levels varied from 36 to <1 during the testing period)

Table 4.19
Water quality data summary of Stagecoach, Nev. first POE GFH run

Parameter	Average	Standard deviation	Maximum value measured	Minimum value measured	Number of samples analyzed	BVs to breakthrough (if applicable)
Arsenic, mg/L	0.005	0.003	0.009	<0.001	6	7,300*
Iron, mg/L	<0.05	0	<0.05	<0.05	1	
Silica, mg/L	24.7	4.1	29.0	4.1	6	

* BVs treated to 0.009 mg/L arsenic breakthrough prior to the column being plugged by iron

Table 4.20
Water quality data summary of Stagecoach, Nev. second POE GFH run (after backwashing)

Parameter	Average	Standard deviation	Maximum value measured	Minimum value measured	Number of samples analyzed	BVs to breakthrough (if applicable)
Arsenic, mg/L	0.009	0.006	0.015	< 0.001	3	4,500*
Iron, mg/L	0.23	0.21	0.58	< 0.05	4	

* BVs treated to 0.010 mg/L arsenic breakthrough after backwashing the POE GFH media

Table 4.21
Source water quality summary from Unity, Maine

Parameter	Result	Number of samples analyzed
pH	8.1	2
Conductivity, μS	220	2
Arsenic, mg/L	0.098	8
Alkalinity, mg/L	85	2
Fluoride, mg/L	<0.4	1
Hardness as $CaCO_3$, mg/L	190	2
Iron, mg/L	0.06	2
Manganese, mg/L	0.06	2
Nitrate as N, mg/L	<0.5	1
Total Phosphorus, mg/L	0.24	1
Sulfate, mg/L	6.3	2
Silica (as SiO_2), mg/L	8.0	2
TDS, mg/L	114	2
Sulfides, mg/L	0.2	historical data

Unity, Maine

A POU RO system and POU Mn-AA system were installed at this site. The POU RO was operated on a 40 minutes on and 80 minutes off cycle for 16 hours, followed by 8 hours of rest. The system was operated without a bladder tank with an approximate flow of 0.1 gpm. The Mn-AA cartridges were tested with a 1.7 min EBCT for the same duration as the RO unit. No preoxidation (to convert the As^{+3} to As^{+5}) was utilized during testing of both POU units.

Groundwater Quality

Summarized in Table 4.21 is the raw water quality for groundwater from Unity, Maine. The arsenic level in the raw water was 0.098 mg/L. Arsenic speciation testing showed that approximately 90% of the arsenic in the Unity groundwater is As^{+3}. The raw water also has low levels of iron (0.06 mg/L), low levels of manganese (0.06 mg/L), high levels of total phosphorus (0.24 mg/L) and trace sulfide levels. The average pH of the raw water was 8.1.

POU RO and Mn-AA Results

The POU RO failed to remove the As^{+3} and was determined to be not suitable for a POU application at this site without preoxidation. Removal of As^{+3} was also not observed during the initial results using the POU Mn-AA cartridge. It was subsequently learned that the POU Mn-AA

Figure 4.22 Misaligned Mn-AA cartridge at Unity, Maine

Table 4.22
Water quality summary of Unity, Maine POU RO effluent

Parameter	Average	Standard deviation	Maximum value measured	Minimum value measured	Number of samples analyzed
pH	8.2	0.01	8.2	8.2	2
Conductivity, μS	432	0	432	431	2
Alkalinity, mg/L	<25	0	<25	<25	1
Arsenic, mg/L	0.075	0.013	0.100	0.053	8
Silica, mg/L	<1	0	<1	<1	4
TDS, mg/L	<15	0	<15	<15	1
Sulfate, mg/L	0.42	0	0.42	0.42	1

cartridge and the housing were misaligned, resulting in the raw water bypassing treatment. Pictures of the misaligned Mn-AA cartridge are presented in Figure 4.22. A centering device was provided by the manufacturer and the cartridge was centered in the housing. Retesting of the Mn-AA cartridge was performed and it was confirmed that it treated up to 1,000 BVs or 640 gallons, with an arsenic level below 10 ppb. Table 4.22 summarizes the effluent sampling data of the POU RO unit. The breakthrough curve for the POU Mn-AA unit is shown in Figure 4.23.

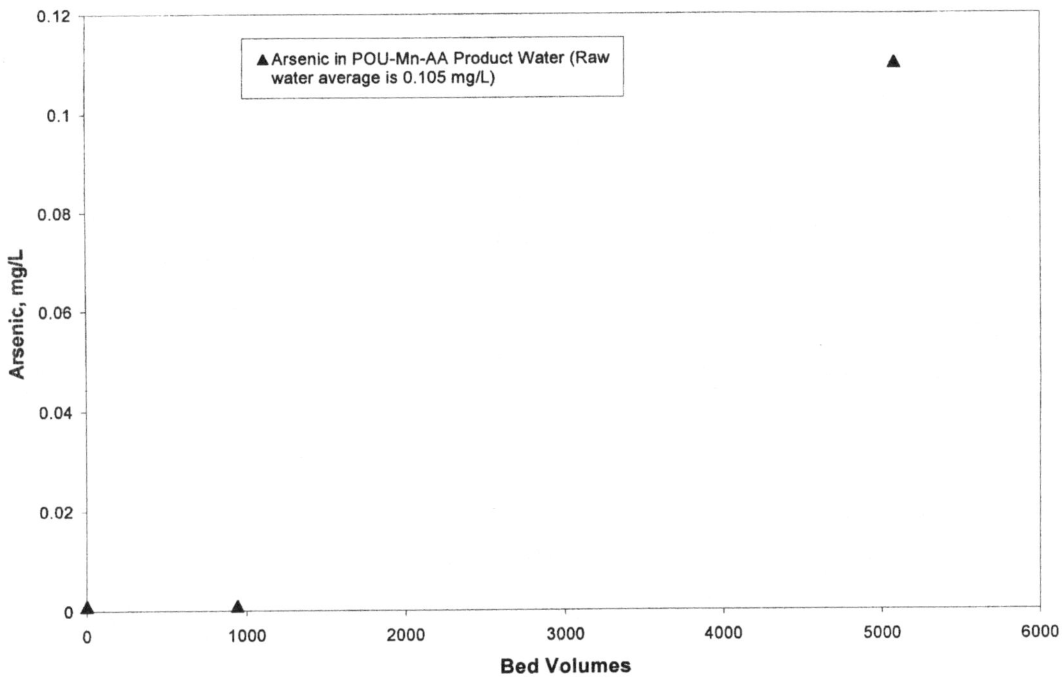

Figure 4.23 Arsenic levels in Unity, Maine POU Mn-AA treated effluent as a function of BVs processed

Carson City, Nev.

A POU Mn-AA system and POU GFH system were installed at this site. The well was operated seasonally and was shut down for several months (testing was discontinued during this time). When the well was returned to service, the testing was completed until breakthrough. The treatment units were operated for approximately 24 hours per day when the well was in operation.

Groundwater Quality

The raw water quality for groundwater from Carson City, Nev. is summarized in Table 4.23. The average arsenic level in the raw water was 0.015 mg/L. The raw water had low levels of manganese, total phosphorus, and iron. The average pH of the raw water was 8.3.

POU Mn-AA and GFH Results

Results for the POU testing at Carson City, Nev. are shown in Tables 4.24 and 4.25 and Figures 4.24 and 4.25. Both the POU units were operated concurrently with the well, with an average operating time of 19 hours per day. The average EBCT was 3.4 minutes. The flow through the contactor was 0.4 gpm. The POU Mn-AA unit treated approximately 5,600 BVs (7,700 gallons treated) with an arsenic level below 10 ppb. The POU GFH column was operated with a 1.7 minute EBCT for the same duration as the POU Mn-AA column. The POU GFH unit treated 22,400 BVs (15,200 gallons treated) before arsenic breakthrough exceeding 10 ppb. Both of the units were initially operated for 10 days before the well was shut down. When the well was

Table 4.23
Source water quality summary from Carson City, Nev.

Parameter	Result	Number of samples analyzed
pH	8.3	2
Conductivity, μS	NA*	NA
Arsenic, mg/L	0.015	2
Alkalinity, mg/L	135	2
Fluoride, mg/L	1.4	2
Hardness as $CaCO_3$, mg/L	190	2
Iron, mg/L	<0.05	2
Manganese, mg/L	NA	NA
Nitrate as N, mg/L	NA	NA
Total Phosphorus, mg/L	0.016	2
Sulfate, mg/L	80.5	2
Silica (as SiO_2), mg/L	17.5	2
TDS, mg/L	304	2
HPCs, CFUs/mL	NA	NA

* not available

Table 4.24
Water quality summary of Carson City, Nev. POU Mn-AA effluent

Parameter	Average	Standard deviation	Maximum value measured	Minimum value measured	Number of samples analyzed	BVs to breakthrough (if applicable)
Arsenic, mg/L	0.007	0.006	0.016	<0.001	9	9,000
Fluoride, mg/L	1.1	0.4	1.4	0.1	8	2,700
Silica, mg/L	14.9	7.5	21.0	1.0	9	1,400

Table 4.25
Water quality summary of Carson City, Nev. POU GFH effluent

Parameter	Average	Standard deviation	Maximum value measured	Minimum value measured	Number of samples analyzed	BVs to breakthrough (if applicable)
Arsenic, mg/L	0.005	0.004	0.012	<0.001	9	26,000
Iron, mg/L	1.6	3.0	7.6	<0.05	9	
Silica, mg/L	17.7	6.5	23.0	1.3	9	5,400

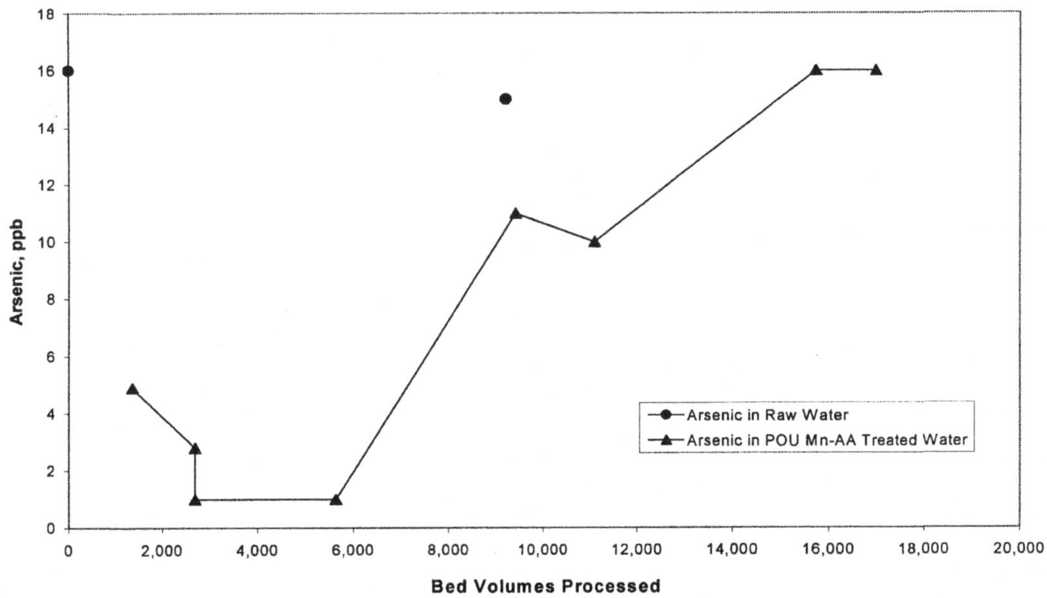

Figure 4.24 Arsenic levels in Carson City, Nev. raw and POU Mn-AA treated effluent as a function of BVs processed

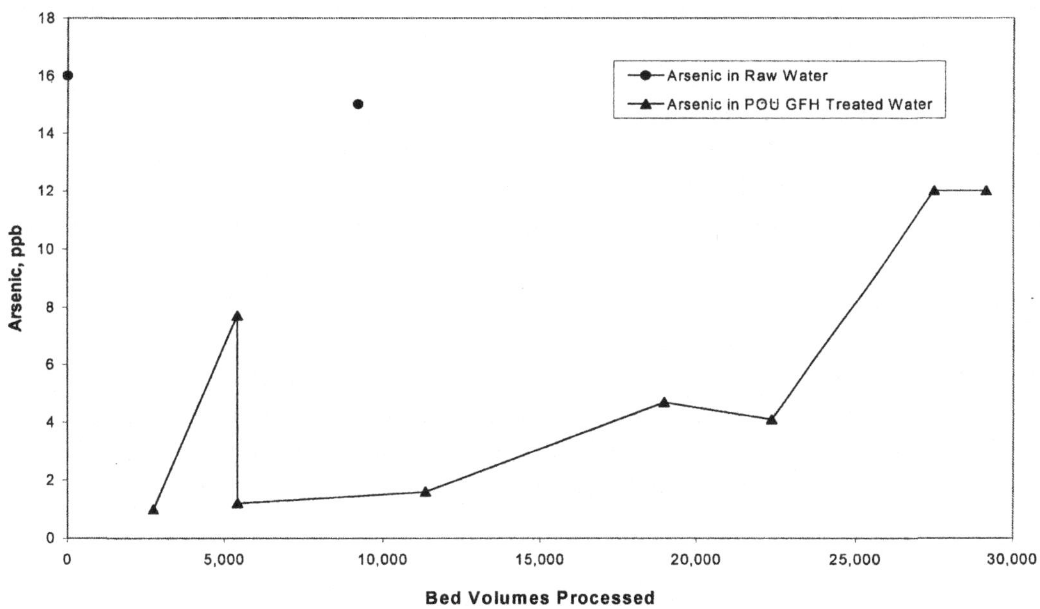

Figure 4.25 Arsenic levels in Carson City, Nev. raw and POU GFH treated effluent as a function of BVs processed

Table 4.26
Source water quality summary for Houston, Texas Terranova West well site

Parameter	Result	Number of samples analyzed
pH	7.6 (average)	6
Conductivity, μS	800	1
Arsenic, mg/L	0.022	1
Alkalinity, mg/L	262	1
Fluoride, mg/L	0.78	1
Hardness as $CaCO_3$, mg/L	88.7	1
Iron, mg/L	0.16	1
Manganese, mg/L	<0.005	1
Nitrate as N, mg/L	<0.10	1
Total Phosphorus, mg/L	<0.10	1
Sulfate, mg/L	8	1
Silica (as SiO_2), mg/L	16.2	1
TDS, mg/L	400	1

returned to service, both cartridges recovered some adsorption capacity, as shown by the lower effluent arsenic levels than the previous results. It is speculated that accumulated arsenic on the surface of the adsorption media migrated into the interstitial pores of the media, resulting in re-availability of surface adsorption sites. This has been observed in other studies (NCS, 2001a).

Houston, Texas

Two POE arsenic treatment systems - Fe-AA with pH adjustment to 6.8, and GFH - were evaluated at this location. The Fe-AA 8-inch diameter column was operated at a flow of 2.0 gpm and a 5 minute EBCT. The column's influent pH was adjusted to 6.8 with sulfuric acid. The GFH 8-inch diameter column was operated at ambient pH with a flow rate of 2.0 gpm and a 2.5 minute EBCT.

Groundwater Quality

Summarized in Table 4.26 is the raw water quality for groundwater from the Houston site. The testing was conducted at a well site in a small municipal utility district within city limits. The raw water was sampled in May 2001. The arsenic level in the raw water was 0.022 mg/L. The raw water also had low levels of manganese and total phosphorus, and detectable levels of iron (0.16 mg/L), which eventually plugged the adsorption units. Initially, this site did not have a prefilter prior to the POE units. The average pH of the raw water was 7.6.

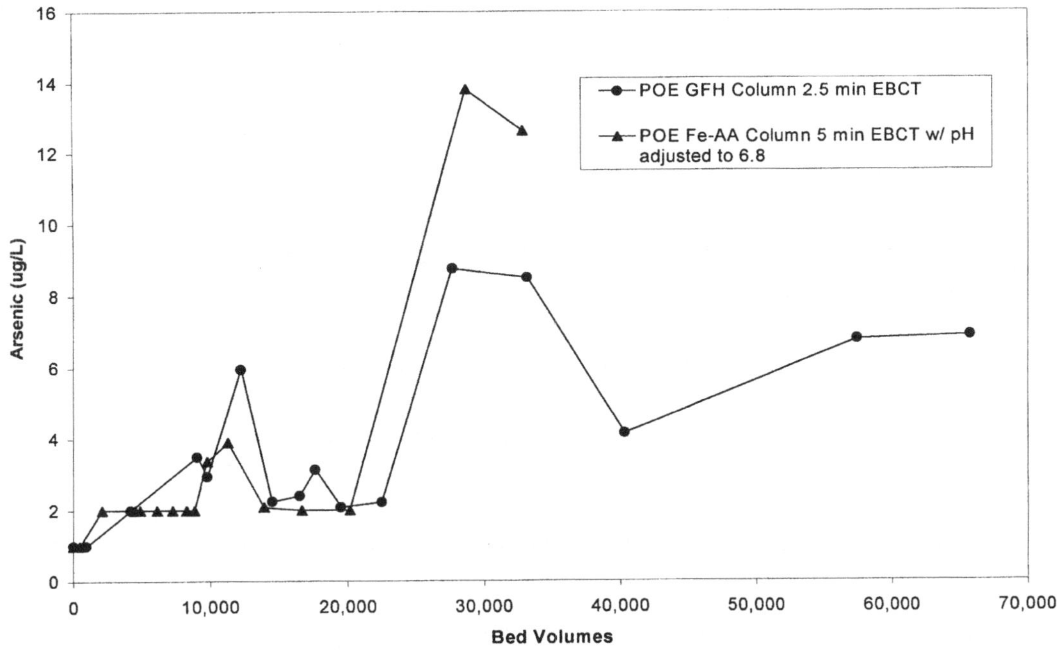

Figure 4.26 Arsenic levels in Houston, Texas POE Fe-AA and POE GFH treated effluents as a function of BVs processed

POE Fe-AA and GFH Results

Plots of arsenic level versus BVs treated and time for the POE Fe-AA and POE GFH systems are shown in Figures 4.26 and 4.27, respectively. Results for the POE Fe-AA and POE GFH units at Houston, Texas are shown in Tables 4.27 and 4.28. The Fe-AA unit processed water for six months, under an almost continuous mode of operation, at a pH of 6.8, until reaching breakthrough at 10 ppb arsenic. At a 5 minute EBCT, this equates to approximately 20,100 BVs (201,400 gallons) treated. The GFH unit was operated continuously for about 2½ months, just before arsenic breakthrough was reached. At this time, the GFH unit was backwashed (late September, 2002). High levels of iron were observed in the backwash streams, as a result of the media being plugged. This caused the effluent arsenic levels to increase sharply from 2.2 ppb to 8.8 ppb (as shown in Figure 4.26). After backwashing was performed, the GFH unit was returned to service, and a 50 micron pre-filter installed. When testing was completed, the GFH unit had treated over 65,800 BVs (328,900 gallons) without reaching arsenic breakthrough. Both of these systems demonstrated the ability to perform as effective POE devices with very long run-times between media replacements (two or more years).

SUMMARY OF FINDINGS FROM POU AND POE PERFORMANCE TESTING

The POU and POE screening evaluation allowed the research team to independently verify manufacturers' test data for arsenic removal. Table 4.29 summarizes the performance of each of the technologies at all of the test sites. The POU adsorption units that were tested at the Tucson Metro Water, Sun City West, and Carson City sites met the criteria for >1,000 gallons treated prior

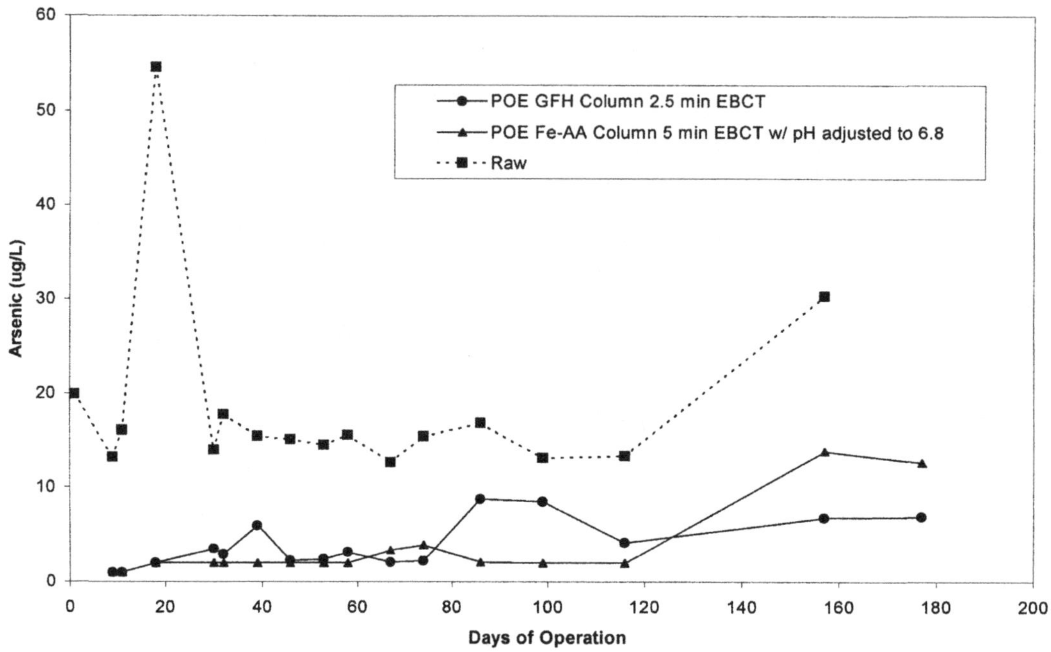

Figure 4.27 Arsenic levels in Houston, Texas POE Fe-AA and POE GFH effluents as a function of time

Table 4.27
Water quality summary of Houston, Texas POE Fe-AA effluent

Parameter	Average	Standard deviation	Maximum value measured	Minimum value measured	Number of samples analyzed	BVs to breakthrough (if applicable)
pH	6.1	0.9	7.0	5.2	3	
Arsenic, mg/L	0.004	0.004	0.014	<0.001	16	20,100
Conductivity, μS	740	0	740	740	1	

Table 4.28
Water quality summary of Houston, Texas POE GFH effluent

Parameter	Average	Standard deviation	Maximum value measured	Minimum value measured	Number of samples analyzed	BVs to breakthrough (if applicable)
pH	7.0	0.8	7.8	6.2	3	
Arsenic, mg/L	0.004	0.003	0.009	0.001	16	>65,800
Conductivity, μS	709	0	709	709	1	

Table 4.29
POU and POE performance summary

Location	Technology	Effluent Si, as SiO_2, mg/L	Effluent pH	Effluent arsenic, mg/L	Gallons treated to 10 ppb breakthrough	Sufficient for 1 yr operation in 5 person home (1000 gal)
Metro Water, Tucson, Ariz.	POU RO	4.8-7.7	6.7-8.8	<0.002	>780	Yes
	POU AA	0.2-15.0	7.4-8.6	<0.002	2,660	Yes
	POE Fe-AA	24-39	7.0-7.7	<0.001-0.010	356,400	Yes
	POE GFH	34-39	7.2-7.7	<0.001-0.006	343,400	Yes
Sun City West, Ariz.	POU RO	0.9-2.3	7.1-8.7	<0.002	>1,300	Yes
	POU AA	<0.1-14.9	7.7-8.4	<0.001-0.025	1,780	Yes
	POU Mn-AA	<0.1-14.2	7.9-8.5	<0.001-0.026	1,780	Yes
	POE Fe-AA	1.3-13.6	7.2-8.5	<0.001-0.022	63,400	Yes
	POE GFH	0.1-15.4	7.2-8.5	<0.001-0.014	368,600	Yes
Stagecoach, Nev.	POE Fe-AA	1.2-26.0	8.0-8.3	<0.001-0.014	34,600	Yes*
	POE GFH	4.1-29.0	8.0-8.3	<0.001-0.009	110,000	Yes
Unity, Maine	POU RO	<1.0	8.2	0.053-0.100	0	Yes†
	POU Mn-AA	7.0-8.5	8.0-8.1	<0.001-0.110	640	Yes
Carson City, Nev.	POU GFH	1.3-23	7.7-8.3	<0.002-0.012	15,200	Yes
	POU Mn-AA	1-21	8.0-9.0	<0.002-0.016	7,700	Yes
Houston, Texas	POE GFH	NA ‡	6.2-7.8	<0.001-0.008	>328,900	Yes
	POE Fe-AA	NA	5.2-7.0	<0.001-0.014	201,450	Yes

* May be applicable with periodic backwashing of the Fe-AA media
† May be applicable with pre-oxidation, prior to treatment
‡ not available

to 10 ppb arsenic breakthrough. The RO units at the Tucson and Sun City West sites treated a lower number of gallons (before reaching arsenic breakthrough) due to the low permeate flux through the system. The RO unit at Unity, Maine failed to remove arsenic and is not suitable as an effective POU application at this site without pretreatment (oxidation of As^{+3} to As^{+5}). The POU units at Carson City, Nev. showed exceptional performance in comparison to other sites. Trace levels of iron in the raw water may have enhanced performance at this site.

All of the POE units demonstrated excellent performance, treating over 50,000 gallons before breakthrough, except the POE Fe-AA unit at Stagecoach, Nev. (34,600 gallons treated). The POE Fe-AA unit at Stagecoach treated <50,000 gallons due to varying levels of iron in the raw water (high levels may plug the column and reduce media adsorption capacity). Periodic backwashing of the Fe-AA column would be required for successful POE application at this site.

This data can be extrapolated with a reasonable level of confidence to home POU systems with lower average to instantaneous flow ratios. Based on the results of the POU field screening evaluation, RO, GFH, AA, and Mn-AA were determined to be the most suitable technologies for further evaluation in the in-home testing program, which is described in Chapter 5.

CHAPTER 5
IN-HOME POU FIELD TESTING PROGRAM

An in-home field testing program that consisted of installing the preferred POU units identified in the screening program was conducted at three small communities whose water supply exceeds the 10 ppb arsenic MCL. The tasks involved in the program included finalizing the list of homes that served as test sites, coordinating with the POU manufacturers and utilities, developing the testing protocol, installing the equipment, monitoring the POU units, performing data analysis, and evaluating performance of the POU systems.

The three sites selected for the study include the Silverbell Water System, Tucson, Ariz; Stagecoach General Improvement District, Nev.; and Sandy Stream Park, Unity, Maine. A description of each system and its groundwater quality is provided below.

TEST SITES

Silverbell Water System, Tucson, Ariz.

The Silverbell Water System serves a community of approximately 75 homes, north of the Tucson city limits near the Town of Marana. The water source for the community is the W1 well, which has an arsenic level of 16-24 ppb, mostly as As (III). The average system flow from the W1 well is 64 gpm. The well water is chlorinated prior to distribution.

Groundwater Quality

Table 5.1 summarizes the raw water quality for the Silverbell West Water System. The raw water was sampled once a week for the first three months of testing. The average arsenic level in the raw water was 0.018 mg/L. The source water had varying iron levels with an average of 0.09 mg/L.

Stagecoach General Improvement District, Stagecoach, Nev.

The Stagecoach General Improvement District consists of two separate systems, each with two wells. The system that participated in this project utilizes the Churchill Rancho North and Churchill Rancho South wells and serves 64 homes. These wells are approximately 500 feet apart and share a common storage tank. Pumping of the wells alternates monthly between the north well and the south well. The instantaneous pumping rate of the two wells is 200 gpm.

Groundwater Quality

The raw water quality for the Stagecoach, Nev. site is summarized in Table 5.2. The arsenic level in the raw water was 0.024 mg/L. The raw water had very low levels of manganese and total phosphorus and varying levels of iron from 0.05 to 2.1 mg/L. With these seasonally varying levels of iron, adsorption units will tend to plug up over long periods of operation. The average pH of the raw water was 8.2.

Table 5.1
Source water quality summary for Silverbell Water System

Parameter	Average value	Maximum value measured	Minimum value measured	Number of samples analyzed
Arsenic, mg/L	0.018	0.024	0.004	18
pH	8.1	8.3	8.0	3
Alkalinity, mg/L	100	109	79	12
Fluoride, mg/L	1.5	2.6	1.0	12
Hardness as $CaCO_3$, mg/L	49	55	39	12
Iron, mg/L	0.09	0.27	<0.02	3
Magnesium, mg/L	0.9	1.1	0.6	12
Manganese, mg/L	0.02	<0.02	<0.02	12
Nitrate, mg/L	0.98	1.00	0.95	12
Orthophosphate, mg/L	<0.2	<0.2	<0.2	12
Sulfate, mg/L	58	66	54	12
Silica, mg/L	12	12	11	12
TDS, mg/L	284	299	264	12
HPC, CFUs/mL	38	210	<1	12

Sandy Stream Park, Unity, Maine

Sandy Stream Park is a small mobile home community near Unity, Maine (northeast of Bangor, ME). The water system serving the community has approximately 40 connections and currently utilizes a 60 gpm central treatment system – pre-oxidation with manganese dioxide media followed by ion exchange. However, the central treatment system is unable to remove arsenic to levels below 10 ppb due to the presence of sulfides, which interfere with arsenite oxidation, in the raw water. Centralized IX treatment has resulted in a cost increase of $42 per month per household. It is estimated that implementation of POU treatment would result in a monthly increase of approximately $22 to $28 per household, depending on the technology selected. The mean annual income of the community is $20,000 per household.

Groundwater Quality

Table 5.3 summarizes the quality of the raw groundwater at the Unity, ME site. The arsenic level in the raw water was 0.098 mg/L, approximately 90 % as arsenic (III). The raw water had low levels of manganese, iron, and total phosphorus, and trace levels of sulfides. The average pH of the raw water was 8.1. The well at the site is currently treated by oxidation with Filox (solid phase oxidizing media) followed by anion exchange. There is no DO in the raw water and 70% of the As (III) is oxidized to As (V) and removed with the anion exchange. The treated effluent that is delivered to the homes has approximately 30 ppb As (III).

Table 5.2
Source water quality summary for Stagecoach General Improvement District

Parameter	Average	Maximum value measured	Minimum value measured	Number of samples analyzed
pH	8.2	8.2	8.1	2
Conductivity, μS	315	315	314	2
Arsenic, mg/L	0.028	0.029	0.027	3
Alkalinity, mg/L	106	113	103	12
Fluoride, mg/L	0.20	0.25	0.13	12
Hardness as $CaCO_3$, mg/L	78	81	74	13
Iron, mg/L	0.10	2.1	0.05	3
Magnesium, mg/L	7.7	8.0	7.2	13
Manganese, mg/L	<0.02	<0.02	<0.02	3
Nitrate as N, mg/L	0.35	0.36	0.34	3
Total Phosphorus, mg/L	<0.2	<0.2	<0.2	3
Sulfate, mg/L	47	49	46	3
Silica (as SiO_2), mg/L	30	31	28	12
TDS, mg/L	250	276	235	12
HPCs, CFUs/mL	100	100	100	1

TESTING PROTOCOL

Sampling Program

A 12-month sampling program was conducted at the Silverbell and Stagecoach water systems to obtain information regarding the performance of the POU units. Due to several unanticipated delays associated with equipment issues and water quality, the sampling program at the Sandy Stream Park in Unity, ME was conducted for only six months. The sampling was conducted by the utility staff with support from the research team. The utility staff were required to go into the homes to collect the samples due to the holding times for analyses such as temperature, pH, total coliform, HPC, etc. The utility staff contacted the homeowners to schedule an appropriate time to collect samples from the units. Members of the research team assisted the utility in implementing the sampling protocol, and periodically accompanied the utility personnel during sampling. Samples from the Silverbell Water System were collected by Tucson Water Department personnel and delivered to the Tucson Water Department water quality laboratory for analyses. Utility personnel from Stagecoach, Nev. and Unity, Maine collected samples from these sites and mailed them to the Tucson Water Department water quality laboratory for analyses. A testing protocol was developed for the in-home field testing program. The testing protocol was designed to include water quality sampling at each home every alternate month. One half of the homes were sampled each month. The sampling protocol is detailed in Table 5.4.

Table 5.3
Source water quality summary for Sandy Stream Park water system

Parameter	Result	Number of samples analyzed
pH	8.1	2
Conductivity, μS	220	2
Arsenic, mg/L	0.098	8
Alkalinity, mg/L	85	2
Fluoride, mg/L	< 0.4	1
Hardness as $CaCO_3$, mg/L	190	2
Iron, mg/L	0.06	2
Manganese, mg/L	0.06	2
Nitrate as N, mg/L	< 0.5	1
Total Phosphorus, mg/L	0.24	1
Sulfate, mg/L	6.3	2
Silica (as SiO_2), mg/L	8.0	2
TDS, mg/L	114	2
Sulfides, mg/L	0.2	historical data

Table 5.4
Twelve-month in-home POU sampling schedule

Locations sampled	Parameters	% of homes sampled and frequency
Raw water	Arsenic, TDS, alkalinity, hardness, temperature, pH, conductivity, HPCs, total coliform, disinfectant residual, sulfate, nitrate, iron, manganese, fluoride, and silica	1 location monthly
Distribution system	Arsenic, HPCs, total coliform, temperature, pH, and disinfectant residual	1 location monthly
Effluent from POU RO	Arsenic, TDS, Alkalinity, Hardness, Temperature, pH, and conductivity	50% of all homes sampled each month (each home samples every other month)
Effluent from Adsorption Unit	Arsenic, pH, temperature	50% of all homes sampled each month (each home samples every other month)
Effluent from POU RO and Adsorption Unit	HPCs, total coliform, disinfectant residual, sulfate, nitrate, iron, manganese, fluoride, and silica	25% of homes sampled on a quarterly basis (each home sampled once during study)

Temperature, pH, and conductivity measurements were taken in the field from the RO units, and pH and temperature were taken from the adsorption units during the monthly sampling events. Hand-held meters were used for these measurements on-site so as not to exceed the holding time for these parameters. Sampling for HPCs (SM 9215), total coliform (SM 9223B), *E. coli* (SM 9223B), disinfectant residual (SM 4500-Cl G), sulfate (EPA 300), nitrate (EPA 300.0), iron (EPA 200.7), manganese (EPA 200.7), fluoride (EPA 300.0), and silica (EPA 200.7) was conducted on a quarterly basis at five of the homes in the Silverbell Water System. Microbiological parameters were not analyzed for the Stagecoach, Nev. and Unity, Maine sites since the sample shipping time exceeded the sample holding time.

INSTALLATION OF POU UNITS

Recruitment of Homes for Test Sites

To solicit participation in the in-home field testing program, informational flyers were distributed to each of the homes in the three participating water systems. In addition, utility personnel and members of the research team visited the homes to discuss the project and inquire about the homeowners' interest in participating. The first 20 homes to sign up were given the opportunity to participate. Twenty homes in the Silverbell Water System and 20 homes in Stagecoach were recruited; however, in the Sandy Stream Park Water System, only four homes were interested in the program.

Equipment Installation

The research team coordinated the installation of the POU units at the test sites with the equipment vendors and utilities. The POU units were set up within the homes by the local vendor. Pictures of the POU RO unit installed in a participating customer's home are presented in Figures 5.1 and 5.2. A picture of an Mn-AA unit and an additional water tap (at the kitchen sink) installed in a participating customer's home are presented in Figure 5.3 and 5.4, respectively. The utility personnel accompanied the vendors during the installation and the subsequent maintenance of the POU units and provided feedback about the level of complexity involved. This also provided an ideal opportunity to determine customer perceptions with respect to maintenance personnel entering homes for replacing membranes and filter cartridges. The research team prepared a post-installation survey (see Appendix E) for the homeowners. To complete the survey, utility personnel and a member of the research team performed site visits at the participating homes. The results of the post-installation survey are discussed in the Customer Perceptions section of this chapter.

The POU devices installed included the Kinetico RO system, Apyron Aquabind MP (Mn-AA), Apyron SP-33 (granular iron media [GIM]), and Aquabind EP (AA) adsorption cartridges that were tested in the field screening portion of this project. The specifications for the units and media are provided in Appendix C. Table 5.5 presents a summary of the technologies installed at the three participating water systems.

Figure 5.1 Under the sink RO unit installed at a participating customer's home

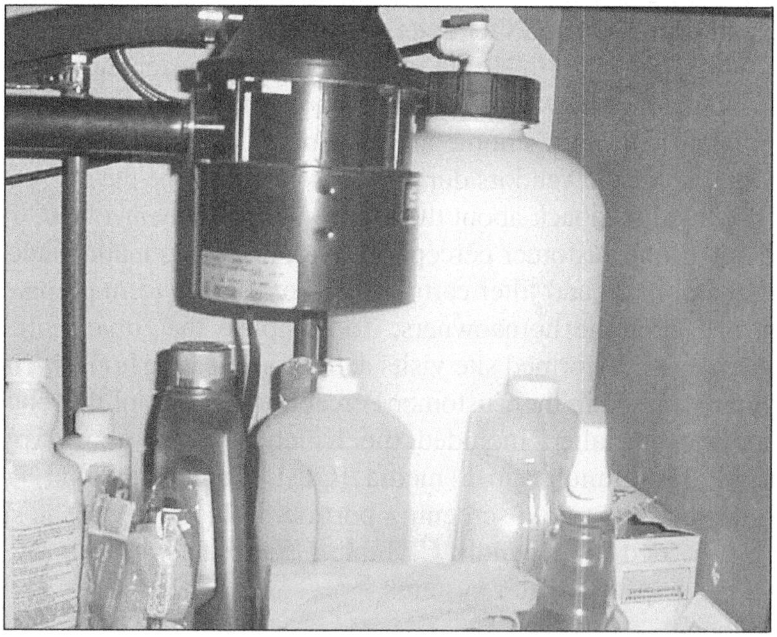

Figure 5.2 Storage tank for the under the sink RO unit

Figure 5.3 Under the sink Mn-AA unit installed in a participating customer's home

Figure 5.4 Additional treated water tap at kitchen sink

Table 5.5
Summary of technologies at in-home test sites

Location	Technologies installed in homes and quantity
Silverbell Water System, Tucson, Ariz.	RO: Osmonics AG membrane - 6 POU: 4.5" × 10" Aquabind MP (Mn-AA) - 7 POU: 4.5" × 10" Aquabind EP (AA) - 2 (subsequently replaced by Mn-AA) POU: 4.5" × 10" SP-33 (GIM) - 5 (subsequently replaced by Mn-AA)
Stagecoach, Nev.	RO: Osmonics AG membrane - 6 POU: 4.5" × 10" Aquabind MP (Mn-AA) - 7 POU: 4.5" × 10" Aquabind EP (AA) - 2 (subsequently replaced by Mn-AA) POU: 4.5" × 10" SP-33 (GIM) - 5 (subsequently replaced by Mn-AA)
Sandy Stream Park, Unity, Maine	POU: 4.5" × 10" Aquabind MP (Mn-AA) - 4

IN-HOME POU FIELD TESTING RESULTS

Silverbell Water System, Tucson, Ariz.

POU Adsorption Results

The adsorption POU units were installed in July 2002. A summary of the quality of the adsorption POU treated water after 3 months is shown in Table 5.6. The initial sample results showed arsenic leakage for all of the adsorption POU units, due to hydraulic short circuiting within the cartridge. In late August 2002, the cartridges were adjusted in an attempt to correct the problem (i.e., the units were readjusted to provide a better seal between the gasket with the housing). Figure 5.5 shows that of the 14 adsorption filters installed in homes in the Silverbell Water System, six of them removed the arsenic to non-detect levels in the effluent. Upon further investigation, it was determined that some of the process flow was bypassing the filter. In addition, the filter housings did not include flow restrictors, causing higher flow rates to occur. This resulted in lower EBCT values in the cartridges and may have impacted arsenic removal performance. The flow rates were regulated subsequently with the installation of needle valves prior to the sink spigot.

Additionally, the internal construction of the adsorption cartridges supplied by the vendor for the in-home installations was different than in the cartridges initially tested in the field screening program. These newer cartridges did not operate correctly due to an improperly designed internal structure, which caused failure. The manufacturer (Apyron) was contacted and made aware of this problem and was directed to supply replacement cartridges with the original design. After investigating the details of the internal construction and seals of the cartridges and conducting discussions with the manufacturer, it was determined that the failures could be attributed to the following reasons:

Table 5.6
Initial sampling results for in-home POU adsorption units with defective cartridges at Silverbell Water System (July-Sept. 2002), see Table 5.7 for representative results

	Arsenic (mg/L)			
POU ID-type	7/25/02	8/14/02	9/4/02	9/11/02
5-EP*	0.021		0.018	
17-EP		0.011		0.012
3-MP†	<0.002		<0.002	
9-MP		<0.002		
10-MP	0.02		<0.002	
11-MP		0.012		0.017
14-MP	0.021		0.018	0.018
15-MP				0.019
20-MP	0.012		0.016	0.018
2-SP-33‡	<0.002		<0.002	
8-SP-33		0.015		<0.002
19-SP-33	0.018		<0.002	
6-SP-33		0.0068		0.0079
13-SP-33		0.011		0.019
Raw	0.018	0.019	0.016	
Distribution System	0.023	0.021	0.019	

* EP = AA adsorption
† MP = Mn-AA adsorption
‡ SP33 = granular iron media

- The in-home cartridges were constructed of a different material (substandard plastic) than the test units used in the POU field screening program. The units could not withstand the pressures and operating condition in the field and many canisters collapsed, resulting in failures.
- The seal between the adsorption canister and the housing was not functioning properly and short-circuiting occurred, resulting in failure of the POU units.
- For this project, the cartridge supplier and housing supplier were two different manufacturers, and the integrity of the whole unit was more difficult to obtain. This is not anticipated to occur in most applications, where one vendor will provide a completed unit with an integral seal.

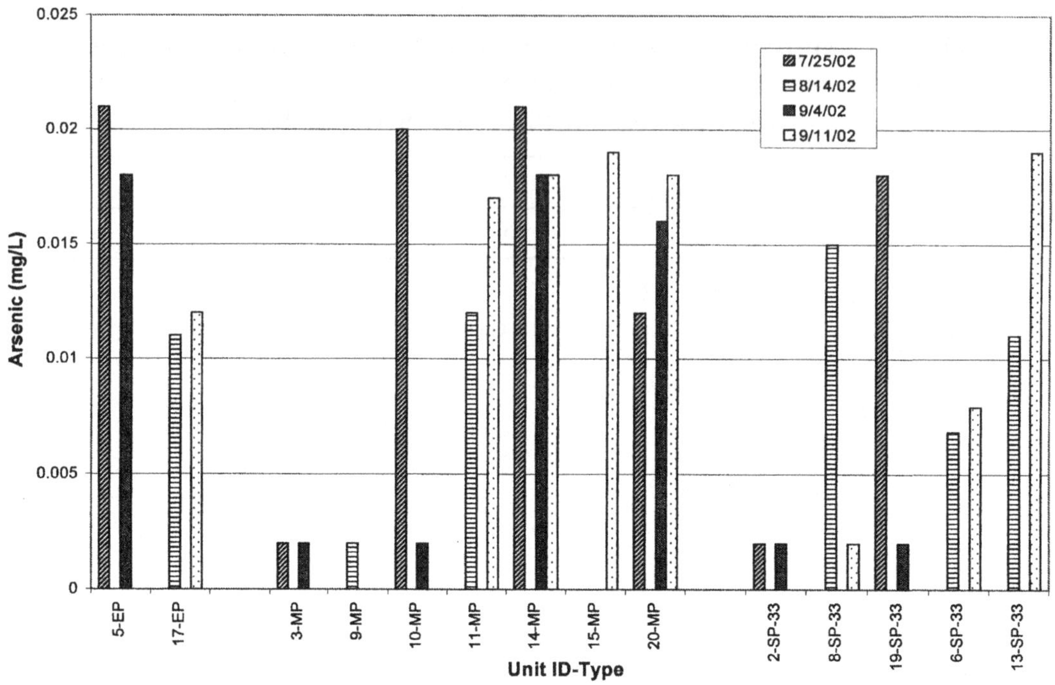

Figure 5.5 Performance of in-home POU adsorption units with defective cartridges at Silverbell Water System (July-Sept. 2002), see Table 5.7 for representative results

In late November 2002, the manufacturer delivered five new GIM SP-33 cartridges using the original design. Testing of each these cartridges was performed at the Tucson well site to verify performance before in-home installation. The results indicated that all five of the replacement cartridges were functioning properly and the design and manufacturing (QA/QC) deficiencies were corrected. The manufacturer subsequently delivered replacement cartridges for the MP (Mn-AA) units. EP (AA) cartridges were not available at that time from the manufacturer and, therefore, were replaced with MP cartridges.

In-home testing of the adsorption units resumed in January 2003. The SP-33 cartridges showed signs of iron media leakage after operating for a few days. The leakage was attributed to fines in the cartridges and design of the internal structure of the cartridge. Due to concerns expressed by the utility and customer complaints, it was decided to replace the SP-33 units with Mn-AA cartridges.

Sampling at the Silverbell Water System resumed in January 2003, after the POU manufacturer deficiencies were corrected and the SP-33 units were replaced with Mn-AA units. Twelve months of sampling was conducted. Results for arsenic, silica, pH, and fluoride for all rounds of sampling are presented in Tables 5.7, 5.8, 5.9, and 5.10, respectively. The average raw water arsenic levels for this community are approximately 0.022 mg/L and range from 0.015-0.027 mg/L. Finished water arsenic levels in 10 of the 12 POU adsorption units were below the detection limit of 0.002 mg/L. Two of the cartridges experienced some arsenic leakage (5-6 ppb [3-5 ppb above the detection limit]). Finished water pH in all the adsorption units ranged from 7.8-8.5 with an average value of 8.0. Finished water silica levels in all the adsorption units ranged from <3-11 mg/L with an average value of 4 mg/L. Finished water fluoride levels in all the adsorption units ranged

Table 5.7
Silverbell Water System in-home POU sampling results for arsenic

Sample ID	POU	Totalizer	7/25/02	8/14/02	8/16/02	9/4/02	9/6/02	9/11/02	9/13/02	1/30/03	2/13/03	02/21/03	2/21/03 Duplicate	2/24/03	3/25/03	4/24/03
5	EP	X	0.021				0.018			<0.002*					<0.002	
17	EP	-		0.011				0.012		<0.002*				<0.002		<0.002
3	MP	X	<0.002				<0.002			0.008		0.006	0.007		0.005	
9	MP	-		<0.002						Discontinued due to leaks						
10	MP	X	0.020			<0.002				<0.002					<0.002	
11	MP	-		0.012				0.017		<0.002				<0.002		<0.002
14	MP	-	0.021			0.018		0.018		<0.002					<0.002	
15	MP	-						0.019		<0.002				<0.002		<0.002
20	MP	-	0.012			0.016		0.018		<0.002					<0.002	
2	SP-33	-	<0.002			<0.002				Discontinued due to leaks						
6	SP-33	X			0.0068				0.0079	<0.002*				<0.002		<0.002
8	SP-33	-		0.015				<0.002		<0.002*				<0.002		<0.002
13	SP-33	X			0.011				0.012		<0.005*			0.004		0.005
19	SP-33	X	0.018			<0.002				<0.002*					<0.002	
1	RO	-	<0.002			<0.002									<0.002	
4	RO	-	<0.002			<0.002									<0.002	
7	RO	-		<0.002										<0.002		<0.002
12	RO	-		<0.002										<0.002		<0.002
16	RO	-		<0.002										<0.002		<0.002
18	RO	-	<0.002			<0.002									<0.002	
W-001A†			0.018	0.019		0.016								0.022	0.021	0.027
SR-262†			0.023	0.021		0.019								0.019	0.022	0.025

(Continued)

Table 5.7 (Continued)

Sample ID	POU	Totalizer	5/27/03	5/30/03	7/29/03	8/20/03	9/3/03	10/16/03	10/17/03	10/22/03	12/11/03
5	EP	X		<0.002	<0.002						
17	EP	-					<0.002		<0.002		<0.002
3	MP	X	0.004		0.006						
9	MP	-									
10	MP	X	<0.002		<0.002						
11	MP	-				<0.002				<0.002	
14	MP	-	<0.002		<0.002						
15	MP	-					<0.002		<0.002		<0.002
20	MP	-	<0.002		<0.002						
2	SP-33	-									
6	SP-33	X				<0.002					<0.002
8	SP-33	-				<0.002				<0.002	<0.002
13	SP-33	X					0.004		0.003		0.004
19	SP-33	X	<0.002		<0.002						
1	RO	-	<0.002		<0.002						
4	RO	-	<0.002		<0.002						
7	RO	-				<0.002		<0.002			<0.002
12	RO	-				<0.002				<0.002	<0.002
16	RO	-				<0.002			<0.002		<0.002
18	RO	-	<0.002		<0.002						
W-001A†			0.017		0.022	0.023			0.015		0.018
SR-262†			0.017		0.021	0.022			0.014		0.020

* Samples collected from this day onwards reflect the new MP cartridge at this location
† Source water

Table 5.8
Silverbell Water System in-home POU sampling results for silica (mg/L)

Sample ID	POU type	8/14/02	9/4/02	2/24/03	3/25/03	4/23/03	5/27/03	7/29/03	8/20/03	9/3/03	10/16/03
5	EP		11.0				<3.0				
17	EP										<3.0
3	MP				<3.0						
9	MP										
10	MP							<3.0			
11	MP	6.9				<3.0					
14	MP							<3.0			
15	MP									<3.0	
20	MP										
2	SP-33										
8	SP-33					<3.0					
19	SP-33										
6	SP-33	5.0		<3.0							
13	SP-33										
1	RO		<5.0		<3.0						
4	RO						<3.0				
7	RO			<3.0							
12	RO										
16	RO										
18	RO								<3.0		

from <0.1-1.5 mg/L with an average value of 0.5 mg/L. Finished water iron and manganese levels in all the adsorption units were below detection limit of 0.02 mg/L (no leaching was observed). Finished water phosphorus levels in all the adsorption units were also below detection limit of 0.2 mg/L.

Six of the Mn-AA adsorption units in the Tucson, Ariz. homes were outfitted with flow totalizers. Table 5.11 shows the totalizer readings recorded between January 2003 and January 2004. More detailed information, including the total volume of water used since the cartridges were replaced in January 2003, total bed volumes processed, total volume used per day, and the total volume used per person per day, are presented in Table 5.12. The flow totalizers recorded between 18 and 650 gallons of total water usage over the duration of 12 months, with an average value of 283 gallons. The average volume of water used was approximately 1 liter/person/day, which is less than the value of 2 liters/person/day used by EPA in setting drinking water standards. A total of approximately 0.8 gallons was used during sample collection. When the homeowners were questioned regarding their POU usage patterns, a variety of responses were given, including the following:

Table 5.9
Silverbell Water System in-home sampling results for pH

Sample ID	POU type	2/24/03	7/29/03	8/20/03	12/11/03
5	EP				
17	EP				8.12
3	MP		7.92		
9	MP				
10	MP		8.11		
11	MP			7.89	
14	MP		8.48		
15	MP				7.84
20	MP		8.18		
2	SP-33				
8	SP-33			7.84	7.75
19	SP-33		7.93		
6	SP-33			7.9	8.06
13	SP-33	8.47			
1	RO		8.16		
4	RO		7.75		
7	RO			6.95	8.39
12	RO			7.09	8.03
16	RO			7.05	7.82
18	RO		7.77		

- Members of the household prefer using the water dispensed from the refrigerator
- The main user of the POU device no longer lives in the home
- The previous filter cartridge (replaced in Jan. 2003) discolored the water

The research team verified with the manufacturer that the flow totalizers were functioning properly. Further information regarding the homeowners' water usage was obtained in the exit survey, which is discussed in the Customer Perceptions section of this chapter.

POU RO Results

As shown in Tables 5.7, 5.8, 5.9, and 5.10, after eleven rounds of sampling (July 2002 to December 2003), all of the RO units were operating efficiently and removed arsenic to below detection limits. Approximately 80-90% removal of TDS and other inorganics was also observed, as shown in Table 5.13. No significant changes were observed in the performance of the units between the various homes.

Table 5.10
Silverbell Water System in-home sampling results for fluoride (mg/L)

Sample ID	POU type	8/14/02	9/4/02	2/24/03*	3/25/03	4/23/03	5/27/03	7/29/03	8/20/03	9/3/03	10/16/03
5	EP		1.3				0.1				
17	EP										0.1
3	MP				0.42						
9	MP										
10	MP							1.3			
11	MP	0.9									
14	MP							0.1			
15	MP									0.11	
20	MP										
2	SP-33										
8	SP-33					0.1					
19	SP-33										
6	SP-33	1.5		0.1							
13	SP-33										
1	RO		0.1		0.1						
4	RO						0.1				
7	RO			0.1							
12	RO										
16	RO	0.1							0.29		
18	RO								0.1		

* samples collected from this day onward reflect the new MP cartridges (adsorption units only)

Table 5.11
Silverbell Water System flow totalizer readings (gallons) from January 2003 to January 2004

Home no.	# of people in home	Jan 03	Feb 03	Mar 03	Apr 03	May 03	July 03	Aug 03	Oct 03	Jan 04
2	3	75		84		92	106			NA*
4	2	320		322		328	335			338
5	3	45	64		84			156		NA
8	5	550		629		673	1076			1200
11	2				31			497	558	636
17	3	116		144		190	248			396
								average totalizer reading		283

* not available

Table 5.12
Silverbell Water System POU water usage data

Home number	# of people in home	POU type	Total volume treated (gallons)	Total volume treated (liters)	Total BVs processed	Total usage volumes (gals/day)	Total usage volumes (mL/day)	Usage volumes per person per day (gal/person/ day)	Usage volumes per person per day (mL/person/ day)
3	3	POU Mn-AA	31	117	56	0.17	651	0.06	0.22
5	2	POU Mn-AA	18	68	33	0.05	189	0.03	0.09
6	3	POU Mn-AA	111	420	202	0.53	1,998	0.18	0.67
10	5	POU Mn-AA	650	2,457	1,182	1.81	6,825	0.36	1.37
13	2	POU Mn-AA	605	2,287	1,100	1.68	6,353	0.84	3.18
19	3	POU Mn-AA	280	1,058	509	0.78	2,940	0.26	0.98
		Average	283	1,068	514	0.84	3,159	0.29	1.08

Table 5.13
Water quality summary of Silverbell POU-RO effluent

Parameter	Maximum value	Minimum value	Average value
Arsenic, mg/L	<0.002	<0.002	<0.002
pH	8.4	7.0	7.7
Alkalinity, mg/L	<20	<20	<20
Fluoride, mg/L	0.29	<0.10	0.13
Hardness as $CaCO_3$, mg/L	<10	<10	<10
Iron, mg/L	<0.020	<0.020	<0.020
Magnesium, mg/L	<0.5	<0.5	<0.5
Manganese, mg/L	<0.020	<0.020	<0.020
Nitrate, mg/L	0.34	<0.25	0.27
Orthophosphate, mg/L	<0.2	<0.2	<0.2
Sulfate, mg/L	<5	<5	<5
TDS, mg/L	22	<10	12

Table 5.14
Initial sampling results for in-home POU adsorption units with defective cartridges at Stagecoach (Aug.-Sept. 2002), see Table 5.15 for representative results

POU ID-type	Arsenic (mg/L) (single data point)
2-EP	<0.002
5-EP	0.0220
1-MP	0.0200
11-MP	<0.002
13-MP	0.0230
16-MP	0.0062
18-MP	0.0098
20-MP	0.0065
Raw	0.0270
Distribution System	0.0210

Stagecoach General Improvement District, Stagecoach, Nev.

POU Adsorption Results

The Stagecoach, Nev. in-home POU units were installed in August 2002. Eight of the adsorption units installed were tested in September 2002, and the arsenic sampling results are shown in Table 5.14. The initial results demonstrated the same problems as observed in the Silverbell (Tucson) homes (e.g., short circuiting due to faulty cartridges and iron release). Figure 5.6 shows that of the eight adsorption cartridge filters tested in the Stagecoach homes, only two of them removed arsenic to non-detect levels in the effluent. The testing was suspended and corrective actions were taken. As with the Silverbell Water System, the manufacturer provided replacement cartridges. However, because of their previous poor performance in the Silverbell homes, the SP-33 (GIM) cartridges were not further evaluated, and the SP-33 and EP cartridges were replaced with MP (Mn-AA) filters.

Sampling at Stagecoach resumed in February 2003, after the POU manufacturer deficiencies were corrected and the SP-33 and EP units were replaced with MP (Mn-AA) units. Eleven rounds of sampling were conducted over 12 months. Results for arsenic, silica, and fluoride for all rounds of sampling are presented in Table 5.15, 5.16, and 5.17, respectively. The raw water arsenic levels for this community ranged from 0.023-0.030 mg/L. Finished water arsenic levels in all but one of the MP POU devices were below the detection limit of 0.002 mg/L. One POU effluent sample in June showed a result of 0.023 mg/L arsenic; subsequent sampling in October 2003 showed a return to an arsenic level below the detection limit of 0.002 mg/L. The exact reason for this aberrant result was not verified (a wrong sample locationb identifier is a possibility). Finished water silica levels in all the adsorption units ranged from location or <3-15 mg/L

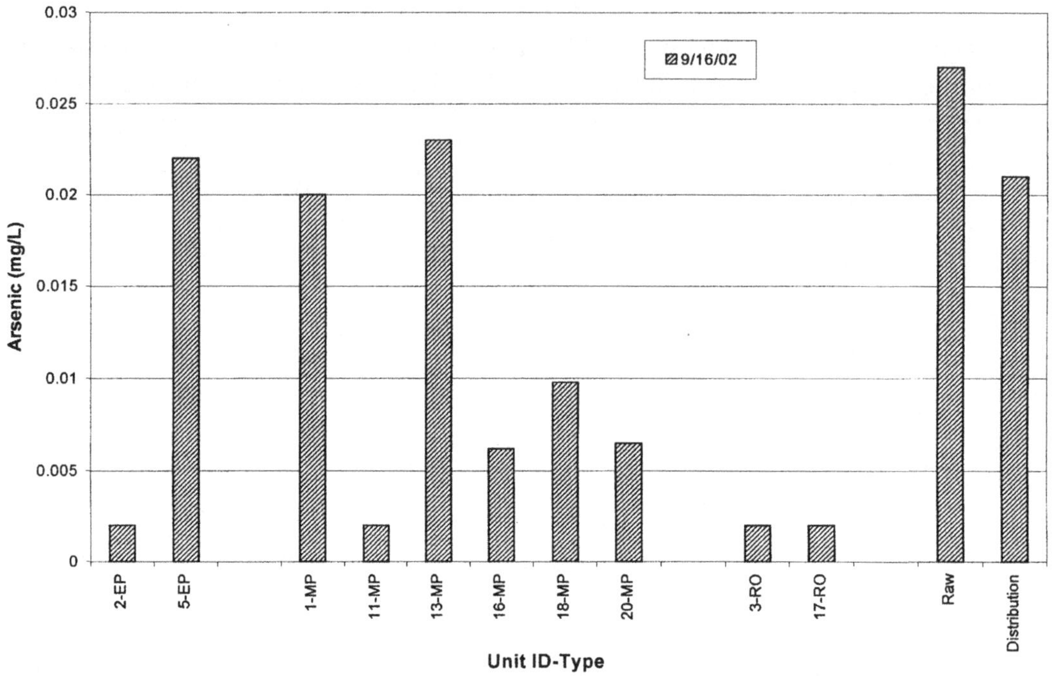

Figure 5.6 Performance of in-home POU adsorption units with defective cartridges at Stagecoach (Aug.-Sept. 2002), see Table 5.15 for representative results

with an average value of 5 mg/L. No pH data is available. Finished water fluoride levels in all the adsorption units ranged from <0.10-0.24 mg/L with an average value of 0.12 mg/L. Finished water iron and manganese levels in all the adsorption units were below detection limits of 0.02 mg/L. Finished water phosphorus levels in all the adsorption units were below detection limits of 0.2 mg/L.

Flow totalizers were installed on four of the MP POU units. Readings of the total number of gallons that passed through the POU units were collected from three of the flow totalizers. The fourth POU unit was removed from the home upon request of the customer. Flow totalizer readings collected in September 2003 and January 2004 are presented in Table 5.18. The totalizer readings ranged from 126 to 235 gallons with an average value of 195 gallons. The per capita usage was 2.3 L/person/day, assuming three persons per household. This value is similar to the value of 2 L/person/day used by EPA in assessing drinking water regulations. Similar to the Silverbell Water System, these readings show that the customers' POU treated water usage varied considerably.

POU RO Results

As shown in Table 5.15, 5.16, and 5.17, after eleven rounds of sampling (September 2002 to December 2003), all of the RO units were operating efficiently and removed arsenic to below detection limits. Approximately 80-90% removal of TDS and other inorganics was also observed, as shown in Table 5.19. No significant changes were observed in the performance of the units between the various homes.

Table 5.15
Stagecoach in-home POU sampling results for arsenic (mg/L)

POU ID	POU type	9/16/02	2/4/03	3/4/03	3/31/03	4/28/03	5/28/03	6/23/03	7/22/03	8/19/03	9/15/03	10/14/03	12/10/03
Raw	n/a		0.025	0.026	0.028	0.027	0.022	0.020	0.030	0.033	0.028	0.024	0.028
Distr.	n/a		0.029	0.028	0.027	0.027	0.023	0.024	0.029	0.030	0.030	0.026	0.026
1	MP			<0.002		<0.002		<0.002		<0.002		<0.002	
23	RO			<0.002		<0.002		<0.002		<0.002		<0.002	
3	RO	<0.002	<0.002		<0.002		<0.002		<0.002		<0.002		<0.002
2	MP			<0.002		<0.002				<0.002		<0.002	
22	RO			<0.002		<0.002		<0.002		<0.002			
21	MP			<0.002		<0.002		0.023				<0.002	
20	MP		<0.002		<0.002		<0.002		<0.002		<0.002		<0.002
19	RO			<0.002		<0.002		<0.002		<0.002		<0.002	
18	MP		<0.002		<0.002		<0.002		<0.002		<0.002		
17	RO	<0.002	<0.002		<0.002		<0.002				<0.002		
26	MP			<0.002		<0.002				<0.002		<0.002	
5	MP		<0.002		<0.002		<0.002		<0.002		<0.002		<0.002
6	MP		<0.002		<0.002		<0.002		<0.002		<0.002		
27	RO			<0.002		<0.002		<0.002		<0.002		<0.002	
12	MP			<0.002		<0.002		<0.002		<0.002		<0.002	
11	MP		<0.002		<0.002		<0.002		<0.002		<0.002		<0.002
10	MP		<0.002		<0.002		<0.002		<0.002		<0.002		<0.002
7	RO		<0.002		<0.002		<0.002		<0.002		<0.002		<0.002
9	MP			<0.002		<0.002		<0.002		<0.002		<0.002	

Table 5.16
Stagecoach in-home POU testing results for silica (mg/L)

POU ID	POU type	9/16/02	3/3/03	4/28/03	5/28/03	6/23/03	7/22/03	8/19/03	9/15/03	12/10/03
1	MP	0.16	<0.10							
2	MP		<0.10							
3	RO	<0.10			<0.10		<0.10		<0.10	<0.10
5	MP									0.24
6	MP									
7	RO				<0.10		<0.10		<0.10	<0.10
9	MP		<0.10							
10	MP									
11	MP									
12	MP		<0.10							
17	RO				<0.10				<0.10	
18	MP									
19	RO		<0.10			<0.10		<0.10		
20	MP									
21	MP		<0.10	0.10						
22	RO		<0.10	<0.10		<0.10		<0.10		
23	RO		<0.10	<0.10		<0.10		<0.10		
26	MP		<0.10	<0.10						
27	RO		<0.10	<0.10		<0.10		<0.10		

Sandy Stream Park, Unity, Maine

POU Adsorption Results

MP (Mn-AA) POU devices were installed in four homes in this community in late January 2004. Results from the first round of samples collected in early February are shown in Table 5.20. An unusually high arsenic spike in the source water caused premature fouling of the media in the MP cartridges; thus, the POU units demonstrated no removal of arsenic. In addition, the water system experienced difficulties with the central treatment system, as the arsenic levels were greater than 200 ppb.

Once the problems were corrected and the central treatment system was able to reduce arsenic levels to <50 ppb, the MP cartridges were replaced and the in-home testing resumed from April 2004. Sampling results for the additional testing are shown in Table 5.21. Samples collected on 4/13/04 are representative of total arsenic, while samples collected on 5/19/04 are representative of As(III). Finished water arsenic levels for As(III) in two of the POU units were slightly less

Table 5.17
Stagecoach in-home POU testing results for fluoride (mg/L)

POU ID	POU type	2/4/03	3/3/03	4/28/03	6/23/03	7/22/03	8/19/03	9/15/03	12/10/03
1	MP		<3.0		<3.0		<3.0		
2	MP		<3.0				<3.0		
3	RO					<3.0		<3.0	<3.0
5	MP	<3.0				<3.0		<3.0	<3.0
6	MP	<3.0				<3.0		<3.0	
7	RO					<3.0		<3.0	<3.0
9	MP		<3.0		3.7		12.0		
10	MP	<3.0				<3.0		<3.0	
11	MP	<3.0				<3.0		<3.0	
12	MP		<3.0		<3.0		<3.0		
17	RO							<3.0	
18	MP	<3.0				<3.0		6.2	
19	RO	<3.0	<3.0		<3.0		<3.0		
20	MP	<3.0				3.6		4.7	
21	MP		<3.0	<3.0	<3.0				
22	RO		<3.0	<3.0	<3.0		<3.0		
23	RO		<3.0	<3.0	<3.0		<3.0		
26	MP		<3.0	<3.0			9.9		
27	RO		<3.0	<3.0	<3.0		<3.0		

Table 5.18
Stagecoach flow totalizer readings

POU ID	Sept 2003 reading (gallons)	Jan 2004 reading (gallons)	Total volume processed (gallons)	Estimated usage (gallons)	Daily usage per person (gal/person/day)	Daily usage per person (L/person/day)	Total bed volumes treated
5	585	810	225	2.1	0.7	2.64	409
10	229	355	126	0.9	0.3	1.13	229
11	725	960	235	2.4	0.8	3.10	427
Average	513	709	195	1.8	0.6	2.3	355

Table 5.19
Stagecoach water quality summary of POU RO effluent

Parameter	Maximum value	Minimum value	Average value
Arsenic, mg/L	<0.002	<0.002	<0.002
Alkalinity, mg/L	<20	<20	<20
Fluoride, mg/L	<0.1	<0.1	<0.1
Hardness as $CaCO_3$, mg/L	<10	<10	<10
Iron, mg/L	<0.020	<0.020	<0.020
Magnesium, mg/L	<0.5	<0.5	<0.5
Manganese, mg/L	<0.020	<0.020	<0.020
Nitrate, mg/L	<0.25	<0.25	<0.25
Orthophosphate, mg/L	<0.2	<0.2	<0.2
Sulfate, mg/L	<5	<5	<5
TDS, mg/L	18	<10	11

Table 5.20
Initial testing results for in-home POU units with defective cartridge at Sandy Stream Park, Unity, Maine

Sample ID	Arsenic, mg/L (single value [2/3/04])
MP1	0.28
MP2	0.27
Raw	0.27

than the raw water arsenic level. One POU unit experienced arsenic leakage of >10 ppb, which may have occurred due to arsenic peaking. Therefore, little or no arsenic removal was observed in the POU Mn-AA units. Based on this limited data, it appears that the POU Mn-AA may not suitable for removal of As(III). The simpler approach would be to oxidize the arsenic at a central location (well) and treat for As(V) with a POU device.

STATISTICAL ASSESSMENT OF POU MONITORING

Based on results of the POU in-home testing performed at Tucson, Ariz. (Silverbell Water System) and Stagecoach, Nev., a statistical analysis was performed to determine optimal monitoring frequencies for POU devices. This analysis will assist in determining appropriate media/membrane replacement frequency prior to arsenic breakthrough and appropriate frequencies for POU monitoring, which provides an opportunity to check equipment performance and address equipment malfunction issues (which may occur due to improper installation, mishandling of the devices, varying raw water quality, etc.). A generalized approach was used based on

Table 5.21
Sampling results for in-home POU units after cartridge replacement at Sandy Stream Park, Unity, Maine

POU ID-type	Arsenic (mg/L) 4/13/04	5/19/04*
MP-1	0.0054	0.0033
MP-2	0.015	0.0072
MP-3	0.0057	0.023
MP-4	0.0073	NA†
Raw	0.0075	0.012
Distribution system	0.0068	0.0085

* results represent As(III)
† not available

Table 5.22
Statistical evaluation of Tucson in-home testing results

		Average arsenic value (mg/L)			
POU type	Total number of homes	Monitoring frequency of 25% homes	Monitoring frequency of 50% homes	Monitoring frequency of 75% homes	Monitoring frequency of 100% homes
POU Mn-AA	12	0.005	0.003	0.003	0.003
POU RO	6	<0.002	<0.002	<0.002	<0.002

the results obtained at the two in-home testing sites. However, the optimal monitoring frequency may vary slightly within each community depending on raw water quality characteristics, seasonal variations in demand, and other site specific issues such as customer acceptance and participation, utility perception, and state regulatory laws.

A summary of the statistical analysis for the POU in-home testing results for Tucson and Stagecoach are presented below. The maximum effluent arsenic level at each in-home site calculated from these single maximums (for the entire testing period) was utilized in the statistical analysis. Average arsenic values for monitoring frequencies of 25%, 50%, 75% and 100% of the total number of homes at the Tucson and Stagecoach in-home sites were determined and are presented in Tables 5.22 and 5.23, respectively. For the 25% frequency, the average of the highest 25% of the single maximum values are shown in Tables 5.22 and 5.23. A similar approach was used for the 50% and 75% values. It is noted that at the Tucson in-home site, the two POU-AA and five POU-SP-33 units were replaced with POU-Mn-AA units at the end of 2002. Therefore, for the purpose of the statistical evaluation, the results of only POU-RO and POU-Mn-AA were used.

Table 5.23
Statistical evaluation of Stagecoach in-home testing results

		Maximum arsenic value (mg/L)			
POU type	Total number of homes	Monitoring frequency of 25% homes	Monitoring frequency of 50% homes	Monitoring frequency of 75% homes	Monitoring frequency of 100% homes
POU Mn-AA	12	<0.002	<0.002	<0.002	<0.002
POU RO	7	<0.002	<0.002	<0.002	<0.002

POU RO Results

The POU-RO units at both the Tucson and Stagecoach sites performed as expected, removing arsenic to below detection limits of <0.002 mg/L even after 12 months of continued usage. Additional sampling at some of the homes at the Tucson site showed that the POU-RO removed arsenic to below detection limits even after 15 months of use. Additionally, no leaks or membrane fouling issues were observed at any of the in-home sites. A survey performed at the end of the study showed that at the Tucson site, three out of six homes who participated in the in-home test opted to continue usage of the POU-RO units, while at Stagecoach, Nev., all seven homes who participated in the in-home testing study opted to keep their POU RO units. Further investigation of these POU units would assist in estimating the exhaustion capacity of the RO membrane.

As shown in Tables 5.22 and 5.23, the average arsenic level is below detection limits for monitoring frequencies of 25%, 50%, 75% and 100% of the total number of homes per year at both Tucson, Ariz. and Stagecoach, Nev. sites. Based on these results, the conservative approach would be a monitoring frequency of one sample per household per year. Since the exact exhaustion capacity of the POU RO units was not determined in this study, the RO membrane is assumed to be replaced every three years. This conservative monitoring frequency approach would ensure periodic maintenance of the units and monitoring for fouling, leaks, mishandling, etc. However, a reduced monitoring frequency of sampling only one-fourth of the total number of households per year may also be acceptable (since all the RO units in the Tucson and Stagecoach areas consistently removed arsenic to below detection limits). This approach would however be accompanied by a membrane replacement frequency of once every two years (regardless of treatment capacity of the POU RO unit).

POU Mn-AA Results

A statistical analysis was performed on the results obtained from the 12 POU Mn-AA units at the Tucson site and from the 12 Mn-AA units at Stagecoach, Nev. after installation of new cartridges in early 2003. All of the POU Mn-AA units at Tucson except two cartridges, which experienced some leakage (5-6 ppb), and all of the POU Mn-AA units at Stagecoach, Nev. removed arsenic to below detection limits of <0.002 mg/L even after 12 months of continued usage. The units performed well without any leakage or malfunctions. A survey performed at the end of the study showed that at the Tucson site, nine out of twelve homes who participated in the in-home test opted to continue usage of the POU-Mn-AA units, while at Stagecoach, all twelve homes who participated in the in-home test opted to continue usage of the POU Mn-AA units. Further investigation of these POU-Mn-AA units would assist in estimating the exhaustion capacity of the Mn-AA media and the BVs to arsenic breakthrough.

As shown in Table 5.22, the average arsenic level is 0.003 mg/L for monitoring frequencies of 50%, 75% and 100% of the total number of homes in the Tucson site, and 0.005 mg/L for a monitoring frequency of 25% of the total number of homes in the Tucson site. As shown in Table 5.23, the arsenic levels are below detection limits for monitoring frequencies of 25%, 50%, 75% and 100% of the total number of homes per year at the Stagecoach site. Based on these results, the conservative approach would be a monitoring frequency of one sample per household per year. This monitoring frequency would ensure periodic maintenance of the units and monitoring for fouling, leaks, mishandling, etc. Since arsenic breakthrough was not detected during the in-home testing, the Mn-AA cartridge is assumed to require replacement every two years. However, a reduced monitoring frequency of sampling only one-fourth of the total number of households per year may be acceptable, as supported by the monitoring date and statistical analysis. This approach would however be accompanied by a more frequent media replacement frequency of once a year (regardless of exhaustion capacity of the POU Mn-AA unit).

EPA Compliance Monitoring

EPA recognizes that statistical representation of the POU units can be used for compliance sampling, which would require only one-fourth to one-ninth of the homes in a community to be sampled each year (USEPA, 2002). Since the POU RO and POU Mn-AA at both Tucson, Ariz. and Stagecoach, Nev. demonstrated excellent performance throughout the test period, the compliance sampling frequency of one-fourth of the homes in a community per year may be applicable. However, the benefits of a higher monitoring frequency of one sample per household per year, with Mn-AA cartridge replacement every two years and RO cartridge replacement every three years, against a lower monitoring frequency of only one-fourth of the homes in a year, with Mn-AA cartridge replacement of once a year and RO cartridge replacement of every two years (regardless of media/membrane exhaustion capacity) should be considered (these details are presented in Chapter 6).

CUSTOMER PERCEPTIONS

As part of the in-home POU field testing program, surveys of the homeowners were conducted to determine customer perception issues with respect to water quality, use of POU units, installation of the POU units, and utility or maintenance personnel entering homes for collecting samples and replacing membranes and filter cartridges. The research team prepared a post-installation survey (see Appendix E) for the homeowners and an exit survey (see Appendix F), which was completed at the end of the study. To complete the surveys, utility personnel and a member of the research team performed site visits at the participating homes.

Post-Installation Survey Results

Following the installation of the POU units, a survey was conducted to evaluate the customers' experience and concerns with the POU installation. In addition, some general consumer information was obtained from the survey. Key observations from the surveys include the following:

- 90% of the volunteers responded favorably to the program with respect to installation and sampling.
- 65% of the volunteers responded that it is very important that they drink arsenic free water.
- 35% of the volunteers would prefer that utility personnel perform the sampling and the installations; the remaining 65% did not indicate a preference between utility personnel and a contractor to perform the tasks.

Exit (End of Study) Survey Results

Prior to the exit survey, utility personnel contacted the homeowners to inform them that the study was nearing completion and that they had the option to keep the POU units, and to schedule a site visit to complete the survey. A survey was conducted to determine the perceptions of the homeowners regarding the overall in-home field testing program, use of POUs and resulting water quality, water consumption rates, and utility or maintenance personnel entering homes for collecting samples and replacing membranes and filter cartridges. In addition, the homeowners provided information regarding leaks or plumbing problems experienced, and whether they planned to keep the POU units. Utility personnel and a member of the research team performed site visits at the participating homes to complete the written survey. Homeowners who were not available for a site visit provided a completed written survey to the utility personnel.

Results from the Silverbell Water System (Tucson) and Stagecoach exit surveys are summarized in the following key observations:

- Most homeowners were highly favorable towards the overall study and the sampling program.
- Most customers thought the POU treated water was better tasting and improved the taste of coffee and other beverages.
- Several homeowners used regular tap water for cooking purposes because the flow rate from the POU tap was too low. Homes in the Tucson site that indicated this experience utilized Mn-AA units, while in Stagecoach, several homeowners experienced low flow rates from RO units.
- Most customers indicated no preference between utility employees and contractors for performance of sampling and maintenance; however, several customers added that they would support the option that is most cost-effective for the community. Several homeowners indicated a preference to perform POU maintenance themselves to reduce costs.
- Many customers experienced problems with small leaks, which were immediately repaired by the installer or utility personnel, from their POUs at the beginning of the sampling program in early 2003. Only two homes (in the Silverbell Water System), however, experienced continued problems with leaking throughout the study. The leaks in these homes were sufficient to cause damage to the cabinets under the sinks, and the researchers provided compensation to the homeowners for repair of the cabinets.

- The majority of participating homes in the Silverbell Water System and all of the participating homes in Stagecoach opted to keep their POU units and accept responsibility for their continued maintenance.

PUBLIC INVOLVEMENT AND COMMUNICATION ISSUES

In order to effectively implement a POU treatment program, it is critical for a utility or water system to develop and follow a public involvement and communication plan. The Silverbell Water System and the Stagecoach General Improvement District were able to effectively involve their customers in the study to obtain the desired number of volunteers. Although two homes in each of the water systems dropped out of the study, the remaining homes remained involved and responded favorably towards the in-home testing program. The opposite was demonstrated by the limited number of volunteers obtained for the Unity, Maine site.

Communication plays an important role in the relationship of a utility with its customers. The Stagecoach General Improvement District maintains a very good relationship with its customers. Frequent communication with the customers aided Stagecoach utility personnel in accessing the homes for POU sampling and maintenance purposes. For those homes with which utility personnel were not able to coordinate convenient sampling times, the utility developed a plan whereby the homeowners collected the samples in the morning and left them on the front porch for the utility to pick up soon thereafter. Utility personnel provided sampling instructions and sample bottles in a bag with the utility's logo and dropped off flyers to remind those customers who had forgotten to collect the samples. In the exit survey, most of the Stagecoach, Nev. customers indicated that the sampling posed no inconvenience and many preferred that utility personnel, rather than a contractor, perform the sampling and maintenance. Utility personnel from Tucson, Ariz., on the other hand, experienced difficulties accessing homes to take samples on several occasions. Problems with coordinating a convenient time with the homeowners for sampling and inability to approach the home due to the presence of unfriendly dogs contributed to these difficulties. Increased communication to make customers aware of such difficulties and the importance of POU sampling and maintenance may help solve these types of problems.

SUMMARY OF RESULTS FROM IN-HOME POU FIELD TESTING PROGRAM

In-home testing of the POU treatment devices (RO and Mn-AA) demonstrated overall excellent performance, removing arsenic to below detection limits of <0.002 mg/L even after 12 months of continued usage. Only two Mn-AA units demonstrated arsenic removals above detections limits; however, despite arsenic leakage of approximately 5-6 ppb, these units still removed arsenic to levels below the 10 ppb MCL. From the statistical assessment of the in-home POU field testing results, it can be concluded that the optimum POU monitoring frequency (conservative approach) for the Silverbell Water System, Tucson, Ariz. and Stagecoach General Improvement District, Nev. may be one sample per household per year, with a media/membrane replacement frequency of once every two years, or until the exhaustion capacity of the POU units is reached. An alternative approach is to sample only one-fourth of the total number of households in a community per year, with a media/membrane replacement frequency of once per year, regardless of exhaustion capacity of the POU units. Optimal monitoring frequencies will vary between communities depending on raw water quality characteristics, seasonal variations in demand, and other site specific issues such as customer acceptance and participation, utility perception, and state regulations.

Although flow totalizer readings indicated that water usage between the participating homes varied considerably (from 338 to 1,200 gallons) over the course of the study, the majority of the homeowners responded favorably with respect to using POU devices and the sampling program. Most homeowners experienced an improvement in the taste of the POU-treated water. Accessing the homes for monitoring and maintenance purposes was generally not a problem for utility personnel. Not surprisingly, there were some incidences where utility personnel experienced problems in coordinating convenient times for sampling and maintenance with the homeowners. The in-home POU field testing program demonstrates the importance of public involvement and communication. In order to effectively manage a POU treatment program, water systems must first get their customers involved. Effective and frequent communication can help water systems maintain a good relationship with their customers, and, in turn, meet POU monitoring and maintenance requirements.

CHAPTER 6
COST ASSESSMENT OF POU AND POE ARSENIC TREATMENT

This chapter presents the findings of a cost assessment of POU and POE treatment options for small water systems, with an emphasis on POU units. A comparison of POU to central treatment is also presented to determine the break-even point between the two options. A comparison of POE and centralized treatment costs is also presented in this chapter. Factors considered in the cost evaluation included:

- Vendor-supplied capital and O&M cost components for POU equipment
- In-home testing results and statistical analysis performed in Chapter 5
- Lease vs. purchase option
- Variable monitoring and maintenance frequencies

A vendor survey to obtain information on POU unit costs and vendor service agreements was performed, the results of which are summarized in the section below. POU treatment costs were estimated based on vendor quotes and media/membrane replacement frequency as determined from the statistical analysis presented in Chapter 5. POE treatment costs are generally similar or higher than centralized treatment costs, therefore a detailed analysis of POE treatment costs was not performed.

POU VENDOR SURVEY

Ten POU equipment suppliers were sent a questionnaire about their POU unit pricing and service agreements. Table 6.1 summarizes the responses from the vendors. Limited information was received from only four vendors, but the information provides a glimpse of the service that is available to consumers. Tables 6.2 and 6.3 summarize the cost quotes received from the vendors for adsorption and RO units, respectively. These costs include pricing for POU cartridges, installation, cartridge replacement, labor, etc. The vendors listed in Table 6.1 do not necessarily correspond to the vendors listed in Tables 6.2 and 6.3.

POU TREATMENT COSTS

Based on the information obtained from the vendor survey and the results of the POU in-home testing program, POU treatment costs were determined. In general, the average price was used in determining the break-even point between POU treatment and central treatment. However, a sensitivity analysis was also performed using minimum and maximum prices to assist utilities in determining the break-even point for their systems.

Cost Evaluation

Capital and O&M costs for POU adsorption units and POU RO units were developed based on the assumptions that the POU systems would be purchased from suppliers with products certified to the ANSI/NSF standards by an ANSI-accredited certifier, and maintenance operations would be performed by the utility. The POU system costs also include costs for monitoring to stay

Table 6.1
Equipment supplier survey responses

Vendor*	Unit	NSF/ANSI certified	Basic service schedule	Service calls response	Rates dependent on location
1	RO unit	NSF	Yearly	24-48 hours	No (charges apply if service required on days other than the predetermined schedule)
2	RO unit	NSF	Yearly	Same day	Yes
3	Undecided	NSF	Contracted out	No response	Depends on leasing company
4	RO unit	NSF	Yearly	24-48 hours	No (cartridge replacement will be performed on a predetermined service schedule)

* Vendor does not correspond with nos. listed in Tables 6.2 and 6.3

Table 6.2
Vendor supplied POU adsorption costs

Vendor	Unit cost*	Installation cost	Replacement cost	Labor cost
1	$210	$100	Cartridge - $90 Pre-filter - $10	Not available
2	$300	$85	$90	$85
3	$350 (including installation)		$130	$50 per service call
4	$125	$50-$150	$70	Not available
5	$150	$50-$150	$35	Not available
Minimum	$125	$50	$70	$50
Maximum	$350	$150	$130	$85
Average	$227	$97	$83	$68

* Adsorption unit typically includes a cartridge and pre-filter

Table 6.3
Vendor supplied POU RO costs

Vendor	Unit cost	Installation cost	Replacement cost	Labor cost
1	$950	$50	Membrane - $96 Pre-filter - $22 Post-filter - $40	$45
2	$600-$1,100	$300-$400	Membrane - $100-$125 Filter cartridges (pre and post) - $20-$25	$90 per service call (includes cartridge change, system sanitization and operation checks)
3	$1,195 (includes sales tax, installation and manufacturer's warranty)	Included in unit cost	Membrane - $119 Pre-filter - $24 Post-filter - $43	$50 per service call (filters to be replaced after 1 year or after 500 gallons, whichever comes first)
4	$210	$50	Filters - $30 Membrane - $48	$50
5	$550	$50-$150	Membrane - $50 Pre-filter - $5 Post-filter - $10	Not available
6	$450	$50-$150	Membrane - $65 Pre-filter - $20 Post-filter - $20	Not available
7	$200	$50-$150	Not available	Not available
8	$200	$50-$150	Membrane - $60 Pre-filter - $18 Post-filter - $18	Not available
Minimum	$200	$50	Membrane - $48 Pre-filter - $5 Post-filter - $10	$45
Maximum	$1,195	$400	Membrane - $125 Pre-filter - $25 Post-filter - $43	$90
Average	$576	$118	Membrane - $80 Pre-filter - $18 Post-filter - $24	$59

in compliance with the Federal and State guidelines. The in-home testing results, presented in Chapter 5, indicate that a media or membrane replacement frequency of >1 year for both the POU Mn-AA and POU RO may be acceptable. The statistical analysis performed in Chapter 5 also indicates that reduced monitoring frequency may be used (as per EPA guidelines). Based on these considerations, the following two scenarios were developed for cost evaluation purposes.

Scenario 1

This scenario is based on the assumption that the frequency of media replacement for a Mn-AA POU unit is once every two years, with a monitoring frequency of one sample per household per year. For the POU RO unit, the frequency of membrane replacement is once every three years, with a monitoring frequency of one sample per household per year. The pre- and post-filter cartridges require replacement once per year (or once 500 gallons have been treated, as per NSF standards), and this would be performed during sample collection. Operator labor charges were assumed at five hours per household per year and managerial charges at two hours per household per year (based on case studies presented in Chapter 5). Scenario 1 is comprised of an increased monitoring frequency with reduced maintenance.

Scenario 2

This scenario is based on the assumption that the adsorption filter cartridges are replaced once per year. The monitoring frequency would be significantly reduced with sample collection from only one-fourth of the total number of households per year. For the POU RO option, the membrane is assumed to be replaced once every two years, and the pre- and post-filters cartridges are assumed to be replaced once per year. Operator labor charges were assumed at 3.5 hours per household per year and managerial charges at two hour per household per year (slight reduction since less maintenance issues are anticipated). Scenario 2 is comprised of a lower monitoring frequency and increased maintenance.

Cost Assumptions

The POU adsorption system was assumed to consist of one adsorption cartridge, with a pre-filter, that can treat approximately 1,000 gallons of water before it has to be replaced. The POU RO system was assumed to be a 3-stage system that includes a pre-filter (for particulate removal), RO membrane, and a post filter (for microbial removal).

Capital costs and operational and maintenance costs for POU adsorption and POU RO units were determined based on the assumptions presented below. Labor costs were developed assuming that periodic maintenance of the POU units is performed by the utility. Therefore, the vendor-supplied labor costs, shown in Tables 6.2 and 6.3, were not used in the cost calculations.

- Average household - three individuals; two liters per day per person consumption; approximately 580 gallons per household per year consumption.
- Installation costs:
  - POU-RO - $118 (average); $50 (low); $400 (high)
  - POU adsorption - $97 (average); $50 (low); $150 (high)

- Equipment:
  - RO unit (includes pre- and post-filters) - $576 (average); $200 (low); $1,195 (high)
  - Adsorption unit - $227 (average); $125 (low); $350 (high)
- Routine maintenance activities including media, membrane, and pre- and post-filter replacement; sample collection, system checks (faulty connections, leaks, etc.) will be performed by the vendor
- Operator labor charges (per hour) - $25
- Managerial charges (per hour) - $50
- Cost of analysis (per sample) - $12
- RO replacement costs (not including labor):
  - Membrane - $80
  - Pre-filter and post-filter - $42
- RO membrane replacement frequency
  - Scenario 1 - once every three years
  - Scenario 2 - once every two years
- Pre- and post-filters (for RO unit) replacement frequency
  - Every 500 gallons (approximately once per year) as per NSF protocol for Scenarios 1 and 2
- Adsorption replacement costs (not including labor) - $85
- Adsorption cartridge replacement frequency
  - Scenario 1 - once every two years
  - Scenario 2 - once per year
- Sampling frequency
  - Scenario 1 - one sample per household per year
  - Scenario 2 - one sample per year from one-fourth of total number of households
- Differential interest rate: 6% (present worth factor of 11.47 over 20 years)
- Analysis period: 20 years.

Cost Calculations

Examples of cost calculations for POU RO treatment and POU adsorption treatment for Scenarios 1 and 2 are shown below:

Household Unit Cost Calculations for Scenario 1

POU RO Annualized Cost	=	(membrane system cost + installation cost)/11.47 + ([membrane, pre-filter and post-filter replacement costs] + labor charges @ 5 hrs/year + managerial charges @ 2 hrs/year + sample analysis cost)
POU RO Annualized Cost (average pricing)	=	($576 + $118)/11.47 + ([$80 × 0.333 + $18 + $24] + [$25 × 5 + $50 × 2] + $12)
	=	$366
	=	$31/month/household

POU RO Annualized Cost (low pricing)	=	($200 + $50)/11.47 + ([$48 × 0.333 + $5 + $10] + [$25 × 5 + $50 × 2] + $12)
	=	$290
	=	$24/month/household
POU RO Annualized Cost (high pricing)	=	($1,195 + $400)/11.47 + ([$125 × 0.333 + $25 + $43] + [$25 × 5 + $50 × 2] + $12)
	=	$486
	=	$40/month/household
POU Adsorption Annualized Cost	=	(adsorption system cost + installation cost)/11.47 + (media replacement cost + labor charges @ 5 hrs/year + managerial charges @ 2 hrs/year + sample analysis cost)
POU Adsorption Annualized Cost (average pricing)	=	($227 + $97)/11.47 + ($85 × 0.5 + $25 × 5 + $50 × 2 + $12)
	=	$308
	=	$26/month/household
POU Adsorption Annualized Cost (low pricing)	=	($125 + $50)/11.47 + ($35 × 0.5 + $25 × 5 + $50 × 2 + $12)
	=	$270
	=	$22/month/household
POU Adsorption Annualized Cost (high pricing)	=	($350 + $150)/11.47 + ($130 × 0.5 + $25 × 5 + $50 × 2 + $12)
	=	$346
	=	$29/month/household

Household Unit Cost Calculations for Scenario 2

POU RO Annualized Cost	=	(membrane system cost + installation cost)/11.47 + ([membrane, pre-filter and post-filter replacement costs] + labor charges @ 3.5 hrs/year + managerial charges @ 2 hrs/year + 25% of sample analysis cost)
POU RO Annualized Cost (average pricing)	=	($576 + $118)/11.47 + ([$80 × 0.5 + $18 + $24] + [$25 × 3.5 + $50 × 2] + 0.25 × $12)
	=	$333
	=	$28/month/household
POU RO Annualized Cost (low pricing)	=	($200 + $50)/11.47 + ([$48 × 0.5 + $5 + $10] + [$25 × 3.5 + $50 × 2] + 0.25 × $12)
	=	$251
	=	$21/month/household
POU RO Annualized Cost (high pricing)	=	($1,195 + $400)/11.47 + ([$125 × 0.5 + $25 + $43] + [$25 × 3.5 + $50 × 2] + 0.25 × $12)
	=	$460
	=	$38/month/household

POU Adsorption Annualized Cost	=	(adsorption system cost + installation cost)/11.47 + (media replacement cost + labor charges @ 3.5 hrs/year + managerial charges @ 2 hrs/year + 25% of sample analysis cost)
POU Adsorption Annualized Cost (average pricing)	=	(\$227 + \$97)/11.47 + (\$85 + \$25 × 3.5 + \$50 × 2 + 0.25 × \$12)
	=	\$301
	=	\$25/month/household
POU Adsorption Annualized Cost (low pricing)	=	(\$125 + \$50)/11.47 + (\$35 + \$25 × 2.5 + \$50 × 1 + 0.25 × \$12)
	=	\$240
	=	\$20/month/household
POU Adsorption Annualized Cost (high pricing)	=	(\$350 + \$150)/11.47 + (\$130 + \$25 × 2.5 + \$50 × 1 + 0.25 × \$12)
	=	\$361
	=	\$30/month/household

A comparison of the POU RO treatment costs (average, low and high values) for Scenarios 1 and 2 is shown in Table 6.4, and a comparison of the average costs is shown in Figure 6.1. As shown in the table, the annualized costs for Scenario 2 are slightly lower than for Scenario 1. The average monthly charges per household for Scenarios 1 and 2 (average values) are \$31 and \$28, respectively. The cost savings obtained by using Scenario 2 are \$3 per household/month (10%).

A comparison of the POU adsorption treatment costs (average, low and high values) for Scenarios 1 and 2 is shown in Table 6.5 and a comparison of the average costs is shown in Figure 6.2. As shown in the table, the annualized costs for Scenario 2 are slightly lower than or similar to those for Scenario 1. The average charges per household per month for Scenarios 1 and 2 are \$26 and \$25, respectively. The cost savings obtained by using Scenario 2 are \$1 per household/month (4%). For both POU RO and POU adsorption treatment, Scenario 1 has slightly higher overall costs. However, the monitoring frequency of one sample per household per year provides the flexibility of scattering the sample collection over the entire community within a year, thereby assuring representative performance of all of the POU units. Therefore, the difference between the two scenarios is a tradeoff between monitoring and maintenance activities. A number of scenarios that compare monitoring vs. maintenance activities exist; however, these two examples are provided in this report.

Purchase/Lease Option

Two options are available to water systems for the procurement of the POU units:

- Purchase the POU units from the vendor; utility is responsible for media cartridge/membrane replacement and other maintenance activities.

Table 6.4
Comparison of POU RO treatment costs for Scenarios 1 and 2

No. of connections		Scenario 1				Scenario 1			
		Capital costs	Annual O&M* costs	Annualized costs	Monthly cost/ household	Capital costs	Annual O&M* costs	Annualized costs	Monthly cost/ household
20	avg.	$13,880	$6,113	$7,320	$31	$13,880	$5,450	$6,660	$28
	low	$5,000	$5,360	$5,800	$24	$5,000	$4,590	$5,030	$21
	high	$31,900	$6,933	$9,710	$40	$31,900	$6,420	$9,200	$38
40	avg.	$27,760	$12,227	$14,650	$31	$27,760	$10,900	$13,320	$28
	low	$10,000	$10,720	$11,590	$24	$10,000	$9,180	$10,050	$21
	high	$63,800	$13,867	$19,430	$40	$63,800	$12,840	$18,400	$38
60	avg.	$41,640	$18,340	$21,970	$31	$41,640	$16,350	$19,980	$28
	low	$15,000	$16,080	$17,390	$24	$15,000	$13,770	$15,080	$21
	high	$95,700	$20,800	$29,140	$40	$95,700	$19,260	$27,600	$38
80	avg.	$55,520	$24,453	$29,290	$31	$55,520	$21,800	$26,640	$28
	low	$20,000	$21,440	$23,180	$24	$20,000	$18,360	$20,100	$21
	high	$127,600	$27,733	$38,860	$40	$127,600	$25,680	$36,800	$38
100	avg.	$69,400	$30,567	$36,620	$31	$69,400	$27,250	$33,300	$28
	low	$25,000	$26,800	$28,980	$24	$25,000	$22,950	$25,130	$21
	high	$159,500	$34,667	$48,570	$40	$159,500	$32,100	$46,010	$38

* Operational and maintenance cost

- Lease the POU units from a vendor; vendor is responsible for all maintenance activities. The utility would still perform sampling and reporting functions, and provide oversight.

Based on vendor quotes, the costs for the lease option are similar to Scenario 1 costs when considering equipment, maintenance, monitoring, reporting, and management activities.

Based on vendor quotes, the costs for leasing the POU units were $20-$25/month for the POU RO and $15 for the POU adsorption units. These costs include equipment only (membrane/media cartridge, pre- and post-filter) and do not include costs for sampling, monitoring and oversight. Therefore, the total lease costs may be only slightly less than or similar to the purchase costs (average and low pricing) even though the operation and maintenance of the POU units is performed by the vendor. The lease option, however, is attractive since the utility would not be responsible for liability and compliance issues, trained operators, staff availability for on-call services, etc. Based on costs and logistics, the lease option is therefore, more viable for procurement of the POU units.

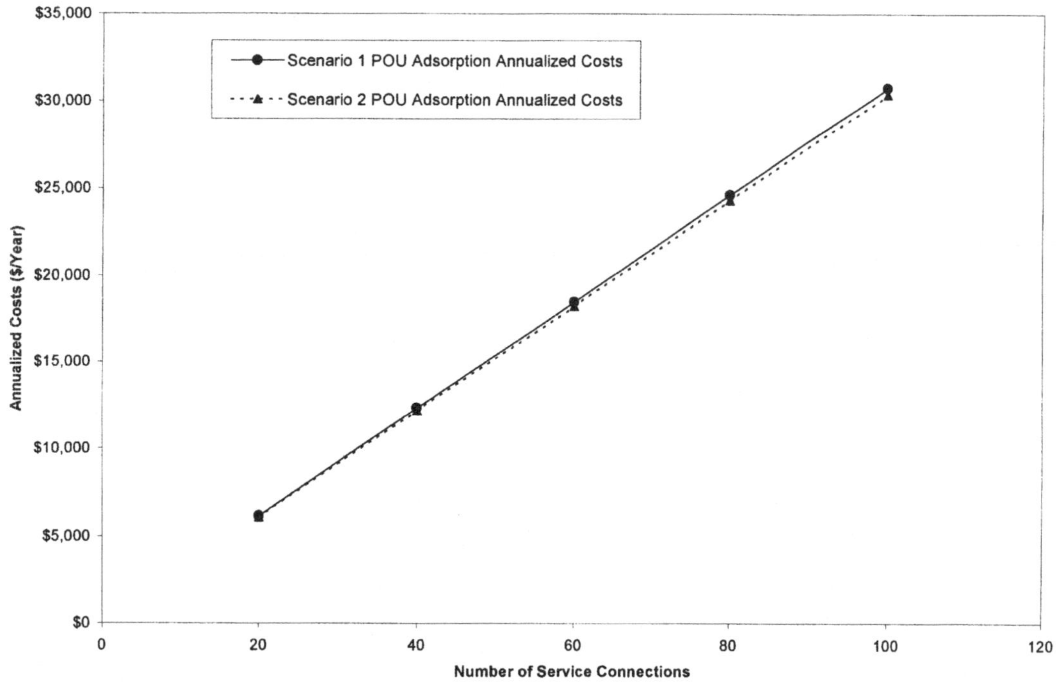

Figure 6.1 Comparison of POU RO treatment costs (average values) for Scenarios 1 and 2

Table 6.5
Comparison of POU adsorption treatment costs for Scenarios 1 and 2

		Scenario 1				Scenario 2			
No. of connections		Capital costs	Annual O&M* costs	Annualized costs	Monthly cost/ household	Capital costs	Annual O&M* costs	Annualized costs	Monthly cost/ household
20	avg.	$6,480	$5,590	$6,150	$26	$6,480	$5,510	$6,070	$25
	low	$3,500	$5,090	$5,400	$22	$3,500	$4,510	$4,820	$20
	high	$10,000	$6,040	$6,910	$29	$10,000	$6,410	$7,280	$30
40	avg.	$12,960	$11,180	$12,310	$26	$12,960	$11,020	$12,150	$25
	low	$7,000	$10,180	$10,790	$22	$7,000	$9,020	$9,630	$20
	high	$20,000	$12,080	$13,820	$29	$20,000	$12,820	$14,560	$30
60	avg.	$19,440	$16,770	$18,460	$26	$19,440	$16,530	$18,220	$25
	low	$10,500	$15,270	$16,190	$22	$10,500	$13,530	$14,450	$20
	high	$30,000	$18,120	$20,740	$29	$30,000	$19,230	$21,850	$30
80	avg.	$25,920	$22,360	$24,620	$26	$25,920	$22,040	$24,300	$25
	low	$14,000	$20,360	$21,580	$22	$14,000	$18,040	$19,260	$20
	high	$40,000	$24,160	$27,650	$29	$40,000	$25,640	$29,130	$30
100	avg.	$32,400	$27,950	$30,770	$26	$32,400	$27,550	$30,370	$25
	low	$17,500	$25,450	$26,980	$22	$17,500	$22,550	$24,080	$20
	high	$50,000	$30,200	$34,560	$29	$50,000	$32,050	$36,410	$30

* Operational and maintenance cost

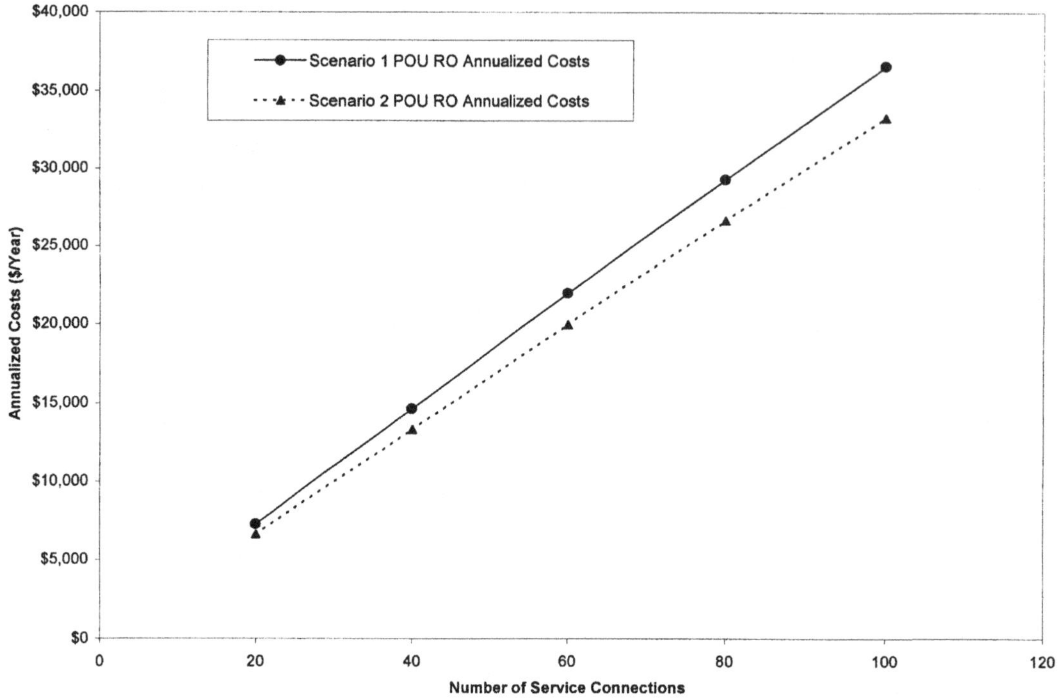

Figure 6.2 Comparison of POU adsorption treatment costs (average values) for Scenarios 1 and 2

POU Costs vs. Central Treatment Costs

A comparison of the costs for POU treatment and central treatment was performed and is presented below. For comparison purposes, the POU treatment costs for Scenario 1 (conservative estimate) were used. Capital and annual O&M costs for centralized treatment using throw-away adsorption media were calculated based on a cost model (NCS, 2000). Comparisons of centralized treatment costs for throw-away adsorption media and average, low, and high POU treatment costs are presented in Tables 6.6, 6.7, and 6.8, respectively, and in Figures 6.3, 6.4 and 6.5, respectively.

Based on average values for POU treatment costs, the POU RO and the POU adsorption treatment costs were significantly lower than centralized treatment costs for systems serving fewer than 50 connections. For these systems, POU RO annualized costs were $18,310, POU adsorption annualized costs were $15,390, and central treatment annualized costs were $28,200 (adsorption using throw-away media). As the number of connections increased, the POU costs also increased gradually, and the breakpoint cost was observed at approximately 120 connections for POU RO and at approximately 200 connections for POU adsorption. The cost per month per household was $31 for POU RO and $26 for POU adsorption, both of which were less than centralized treatment cost per month per household, for less than 120 and 200 connections respectively. The annual O&M costs for the POU RO and POU adsorption units were higher than for central treatment, for systems serving greater than 50 and 60 connections, respectively. These higher annual O&M costs can be attributed to an increase in administrative, monitoring, and membrane/media cartridge and pre- and post-filter replacement costs. The capital costs of the POU units, however, were consistently lower than central treatment capital costs. From an overall cost evaluation, POU treatment is cost-effective when compared to centralized treatment and may be viable for very small systems.

Table 6.6
Comparison of centralized and Scenario 1 POU treatment costs (average values)

				POU RO				POU adsorption				Central treatment (Fe-AA)			
No. of connections	No. of people served	Flow (gpd)	Flow (MGD)	POU capital costs	Annual O&M* costs	Annualized costs	Monthly cost/ household	Capital costs	Annual O&M* costs	Annualized costs	Monthly cost/ household	Capital costs	Annual O&M* costs	Annualized costs	Monthly cost/ household
20	60	9,000	0.02	$13,880	$6,113	$7,320	$31	$6,480	$5,590	$6,150	$26	$139,600	$9,400	$21,600	$90
30	90	13,500	0.03	$20,820	$9,170	$10,990	$31	$9,720	$8,390	$9,240	$26	$147,100	$11,000	$23,800	$66
40	120	18,000	0.04	$27,760	$12,227	$14,650	$31	$12,960	$11,180	$12,310	$26	$154,700	$12,500	$26,000	$54
50	150	22,500	0.05	$34,700	$15,283	$18,310	$31	$16,200	$13,980	$15,390	$26	$162,200	$14,100	$28,200	$47
60	180	27,000	0.05	$41,640	$18,340	$21,970	$31	$19,440	$16,770	$18,460	$26	$169,800	$15,700	$30,500	$42
70	210	31,500	0.06	$48,580	$21,397	$25,630	$31	$22,680	$19,570	$21,550	$26	$177,300	$17,300	$32,800	$39
80	240	36,000	0.07	$55,520	$24,453	$29,290	$31	$25,920	$22,360	$24,620	$26	$184,900	$18,900	$35,000	$36
90	270	40,500	0.08	$62,460	$27,510	$32,960	$31	$29,160	$25,160	$27,700	$26	$192,400	$20,500	$37,300	$35
100	300	45,000	0.09	$69,400	$30,567	$36,620	$31	$32,400	$27,950	$30,770	$26	$200,000	$22,100	$39,500	$33
110	330	49,500	0.10	$76,340	$33,623	$40,280	$31	$35,640	$30,750	$33,860	$26	$207,500	$23,700	$41,800	$32
120	360	54,000	0.11	$83,280	$36,680	$43,940	$31	$38,880	$33,540	$36,930	$26	$215,100	$25,300	$44,100	$31
130	390	58,500	0.12	$90,220	$39,737	$47,600	$31	$42,120	$36,340	$40,010	$26	$222,600	$26,900	$46,300	$30
140	420	63,000	0.13	$97,160	$42,793	$51,260	$31	$45,360	$39,130	$43,080	$26	$230,200	$28,500	$48,600	$29
150	450	67,500	0.14	$104,100	$45,850	$54,930	$31	$48,600	$41,930	$46,170	$26	$237,700	$30,100	$50,800	$28
160	480	72,000	0.14	$111,040	$48,907	$58,590	$31	$51,840	$44,720	$49,240	$26	$245,300	$31,700	$53,100	$28
170	510	76,500	0.15	$117,980	$51,963	$62,250	$31	$55,080	$47,520	$52,320	$26	$252,800	$33,300	$55,300	$27
180	540	81,000	0.16	$124,920	$55,020	$65,910	$31	$58,320	$50,310	$55,390	$26	$260,400	$34,900	$57,600	$27
190	570	85,500	0.17	$131,860	$58,077	$69,570	$31	$61,560	$53,110	$58,480	$26	$267,900	$36,500	$59,900	$26
200	600	90,000	0.18	$138,800	$61,133	$73,230	$31	$64,800	$55,900	$61,550	$26	$275,500	$38,100	$62,100	$26
210	630	94,500	0.19	$145,740	$64,190	$76,900	$31	$68,040	$58,700	$64,630	$26	$283,000	$39,700	$64,400	$26
220	660	99,000	0.20	$152,680	$67,247	$80,560	$31	$71,280	$61,490	$67,700	$26	$290,600	$41,300	$66,600	$25
230	690	103,500	0.21	$159,620	$70,303	$84,220	$31	$74,520	$64,290	$70,790	$26	$298,100	$42,800	$68,800	$25
240	720	108,000	0.22	$166,560	$73,360	$87,880	$31	$77,760	$67,080	$73,860	$26	$305,700	$44,400	$71,100	$25
250	750	112,500	0.23	$173,500	$76,417	$91,540	$31	$81,000	$69,880	$76,940	$26	$313,200	$46,000	$73,300	$24
260	780	117,000	0.23	$180,440	$79,473	$95,200	$31	$84,240	$72,670	$80,010	$26	$320,800	$47,600	$75,600	$24
270	810	121,500	0.24	$187,380	$82,530	$98,870	$31	$87,480	$75,470	$83,100	$26	$328,300	$49,200	$77,800	$24
280	840	126,000	0.25	$194,320	$85,587	$102,530	$31	$90,720	$78,260	$86,170	$26	$335,900	$50,800	$80,100	$24
290	870	130,500	0.26	$201,260	$88,643	$106,190	$31	$93,960	$81,060	$89,250	$26	$343,400	$52,400	$82,300	$24
300	900	135,000	0.27	$208,200	$91,700	$109,850	$31	$97,200	$83,850	$92,320	$26	$351,000	$54,000	$84,600	$24

* Operation and maintenance cost

Table 6.7
Comparison of centralized and Scenario 1 POU treatment costs (low range values)

				POU RO				POU adsorption				Central treatment (Fe-AA)			
No. of connections	No. of people served	Flow (gpd)	Flow (MGD)	POU capital costs	Annual O&M* costs	Annualized costs	Monthly cost/ household	Capital costs	Annual O&M* costs	Annualized costs	Monthly cost/ household	Capital costs	Annual O&M* costs	Annualized costs	Monthly cost/ household
20	60	9,000	0.02	$5,000	$5,360	$5,800	$24	$3,500	$5,090	$5,400	$23	$139,600	$9,400	$21,600	$90
30	90	13,500	0.03	$7,500	$8,040	$8,690	$24	$5,250	$7,635	$8,090	$22	$147,100	$11,000	$23,800	$66
40	120	18,000	0.04	$10,000	$10,720	$11,590	$24	$7,000	$10,180	$10,790	$22	$154,700	$12,500	$26,000	$54
50	150	22,500	0.05	$12,500	$13,400	$14,490	$24	$8,750	$12,725	$13,490	$22	$162,200	$14,100	$28,200	$47
60	180	27,000	0.05	$15,000	$16,080	$17,390	$24	$10,500	$15,270	$16,190	$22	$169,800	$15,700	$30,500	$42
70	210	31,500	0.06	$17,500	$18,760	$20,290	$24	$12,250	$17,815	$18,880	$22	$177,300	$17,300	$32,800	$39
80	240	36,000	0.07	$20,000	$21,440	$23,180	$24	$14,000	$20,360	$21,580	$22	$184,900	$18,900	$35,000	$36
90	270	40,500	0.08	$22,500	$24,120	$26,080	$24	$15,750	$22,905	$24,280	$22	$192,400	$20,500	$37,300	$35
100	300	45,000	0.09	$25,000	$26,800	$28,980	$24	$17,500	$25,450	$26,980	$22	$200,000	$22,100	$39,500	$33
110	330	49,500	0.10	$27,500	$29,480	$31,880	$24	$19,250	$27,995	$29,670	$22	$207,500	$23,700	$41,800	$32
120	360	54,000	0.11	$30,000	$32,160	$34,780	$24	$21,000	$30,540	$32,370	$22	$215,100	$25,300	$44,100	$31
130	390	58,500	0.12	$32,500	$34,840	$37,670	$24	$22,750	$33,085	$35,070	$22	$222,600	$26,900	$46,300	$30
140	420	63,000	0.13	$35,000	$37,520	$40,570	$24	$24,500	$35,630	$37,770	$22	$230,200	$28,500	$48,600	$29
150	450	67,500	0.14	$37,500	$40,200	$43,470	$24	$26,250	$38,175	$40,460	$22	$237,700	$30,100	$50,800	$28
160	480	72,000	0.14	$40,000	$42,880	$46,370	$24	$28,000	$40,720	$43,160	$22	$245,300	$31,700	$53,100	$28
170	510	76,500	0.15	$42,500	$45,560	$49,270	$24	$29,750	$43,265	$45,860	$22	$252,800	$33,300	$55,300	$27
180	540	81,000	0.16	$45,000	$48,240	$52,160	$24	$31,500	$45,810	$48,560	$22	$260,400	$34,900	$57,600	$27
190	570	85,500	0.17	$47,500	$50,920	$55,060	$24	$33,250	$48,355	$51,250	$22	$267,900	$36,500	$59,900	$26
200	600	90,000	0.18	$50,000	$53,600	$57,960	$24	$35,000	$50,900	$53,950	$22	$275,500	$38,100	$62,100	$26
210	630	94,500	0.19	$52,500	$56,280	$60,860	$24	$36,750	$53,445	$56,650	$22	$283,000	$39,700	$64,400	$26
220	660	99,000	0.20	$55,000	$58,960	$63,760	$24	$38,500	$55,990	$59,350	$22	$290,600	$41,300	$66,600	$25
230	690	103,500	0.21	$57,500	$61,640	$66,650	$24	$40,250	$58,535	$62,040	$22	$298,100	$42,800	$68,800	$25
240	720	108,000	0.22	$60,000	$64,320	$69,550	$24	$42,000	$61,080	$64,740	$22	$305,700	$44,400	$71,100	$25
250	750	112,500	0.23	$62,500	$67,000	$72,450	$24	$43,750	$63,625	$67,440	$22	$313,200	$46,000	$73,300	$24
260	780	117,000	0.23	$65,000	$69,680	$75,350	$24	$45,500	$66,170	$70,140	$22	$320,800	$47,600	$75,600	$24
270	810	121,500	0.24	$67,500	$72,360	$78,240	$24	$47,250	$68,715	$72,830	$22	$328,300	$49,200	$77,800	$24
280	840	126,000	0.25	$70,000	$75,040	$81,140	$24	$49,000	$71,260	$75,530	$22	$335,900	$50,800	$80,100	$24
290	870	130,500	0.26	$72,500	$77,720	$84,040	$24	$50,750	$73,805	$78,230	$22	$343,400	$52,400	$82,300	$24
300	900	135,000	0.27	$75,000	$80,400	$86,940	$24	$52,500	$76,350	$80,930	$22	$351,000	$54,000	$84,600	$24
320	960	144,000	0.29	$80,000	$85,760	$92,730	$24	$56,000	$81,440	$86,320	$22	$366,100	$57,200	$89,100	$23
340	1020	153,000	0.31	$85,000	$91,120	$98,530	$24	$59,500	$86,530	$91,720	$22	$381,200	$60,400	$93,600	$23
360	1080	162,000	0.32	$90,000	$96,480	$104,330	$24	$63,000	$91,620	$97,110	$22	$396,300	$63,600	$98,200	$23
380	1140	171,000	0.34	$95,000	$101,840	$110,120	$24	$66,500	$96,710	$102,510	$22	$411,400	$66,800	$102,700	$23
400	1200	180,000	0.36	$100,000	$107,200	$115,920	$24	$70,000	$101,800	$107,900	$22	$426,500	$70,000	$107,200	$22

* Operation and maintenance cost

Table 6.8
Comparison of centralized and Scenario 1 POU treatment costs (high range values)

				POU RO				POU adsorption				Central treatment (Fe-AA)			
No. of connections	No. of people served	Flow (gpd)	Flow (MGD)	POU capital costs	Annual O&M* costs	Annualized costs	Monthly cost/ household	Capital costs	Annual O&M* costs	Annualized costs	Monthly cost/ household	Capital costs	Annual O&M costs	Annualized costs	Monthly cost/ household
20	60	9.000	0.02	$31,900	$6,933	$9,710	$40	$10,000	$6,040	$6,910	$29	$139,600	$9,400	$21,600	$90
30	90	13.500	0.03	$47,850	$10,400	$14,570	$40	$15,000	$9,060	$10,370	$29	$147,100	$11,000	$23,800	$66
40	120	18,000	0.04	$63,800	$13,867	$19,430	$40	$20,000	$12,080	$13,820	$29	$154,700	$12,500	$26,000	$54
50	150	22,500	0.05	$79,750	$17,333	$24,290	$40	$25,000	$15,100	$17,280	$29	$162,200	$14,100	$28,200	$47
60	180	27,000	0.05	$95,700	$20,800	$29,140	$40	$30,000	$18,123	$20,740	$29	$169,800	$15,700	$30,500	$42
70	210	31,500	0.06	$111,650	$24,267	$34,000	$40	$35,000	$21,140	$24,190	$29	$177,300	$17,300	$32,800	$39
80	240	36,000	0.07	$127,600	$27,733	$38,860	$40	$40,000	$24,160	$27,650	$29	$184,900	$18,900	$35,000	$36
90	270	40,500	0.08	$143,550	$31,200	$43,720	$40	$45,000	$27,180	$31,100	$29	$192,400	$20,500	$37,300	$35
100	300	45,000	0.09	$159,500	$34,667	$48,570	$40	$50,000	$30,200	$34,560	$29	$200,000	$22,100	$39,500	$33
110	330	49,500	0.10	$175,450	$38,133	$53,430	$40	$55,000	$33,220	$38,020	$29	$207,500	$23,700	$41,800	$32
120	360	54,000	0.11	$191,400	$41,600	$58,290	$40	$60,000	$36,240	$41,470	$29	$215,100	$25,300	$44,100	$31
130	390	58,500	0.12	$207,350	$45,067	$63,143	$40	$65,000	$39,260	$44,930	$29	$222,600	$26,900	$46,300	$30
140	420	63,000	0 13	$223,300	$48,533	$68,000	$40	$70,000	$42,280	$48,380	$29	$230,200	$28,500	$48,600	$29
150	450	67,500	0 14	$239,250	$52,000	$72,860	$40	$75,000	$45,300	$51,840	$29	$237,700	$30,100	$50,800	$28
160	480	72,000	0.14	$255,200	$55,467	$77,720	$40	$80,000	$48,320	$55,290	$29	$245,300	$31,700	$53,100	$28
170	510	76,500	0.15	$271,150	$58,933	$82,570	$40	$85,000	$51,340	$58,750	$29	$252,800	$33,300	$55,300	$27
180	540	81,000	0.16	$287,100	$62,400	$87,430	$40	$90,000	$54,360	$62,210	$29	$260,400	$34,900	$57,600	$27
190	570	85,500	0.17	$303,050	$65,867	$92,290	$43	$95,000	$57,380	$65,660	$29	$267,900	$36,500	$59,900	$26
200	600	90,000	0 18	$319,000	$69,333	$97,150	$40	$100,000	$60,400	$69,120	$29	$275,500	$38,100	$62,100	$26
210	630	94,500	0.19	$334,950	$72,800	$102,000	$40	$105,000	$63,420	$72,570	$29	$283,000	$39,700	$64,400	$26
220	660	99,000	0.20	$350,900	$76,267	$106,860	$40	$110,000	$66,440	$76,030	$29	$290,600	$41,300	$66,600	$25
230	690	103,500	0.21	$366,850	$79,733	$111,720	$40	$115,000	$69,460	$79,490	$29	$298,100	$42,800	$68,800	$25
240	720	108,000	0.22	$382,800	$83,200	$116,570	$40	$120,000	$72,480	$82,940	$29	$305,700	$44,400	$71,100	$25
250	750	112,500	0.23	$398,750	$86,667	$121,430	$40	$125,000	$75,500	$86,400	$29	$313,200	$46,000	$73,300	$24
260	780	117,000	0.23	$414,700	$90,133	$126,290	$40	$130,000	$78,520	$89,850	$29	$320,800	$47,600	$75,600	$24
270	810	121,500	0.24	$430,650	$93,600	$131,150	$40	$135,000	$81,540	$93,310	$29	$328,300	$49,200	$77,800	$24
280	840	126,000	0.25	$446,600	$97,067	$136,000	$40	$140,000	$84,560	$96,770	$29	$335,900	$50,800	$80,100	$24
290	870	130,500	0.26	$462,550	$190,533	$140,860	$40	$145,000	$87,580	$100,220	$29	$343,400	$52,400	$82,300	$24
300	900	135,000	0.27	$478,500	$104,000	$145,720	$40	$150,000	$90,600	$133,680	$29	$351,000	$54,000	$84,600	$24

* Operation and maintenance cost

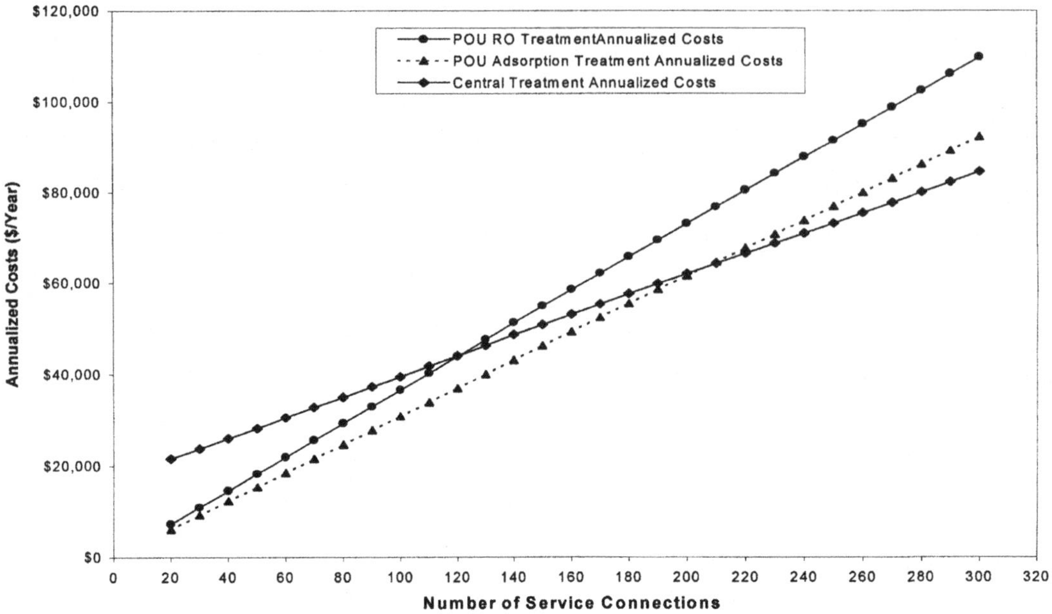

Figure 6.3 Comparison of central and Scenario 1 POU annualized treatment costs (average values)

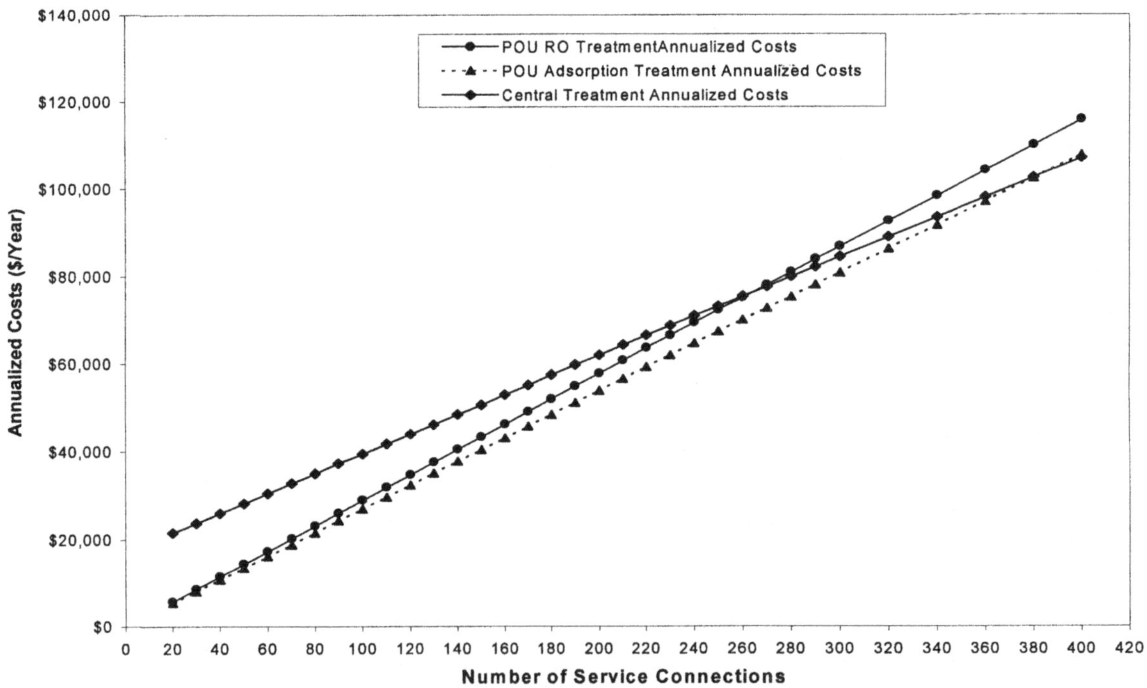

Figure 6.4 Comparison of central and Scenario 1 POU annualized treatment costs (low range values)

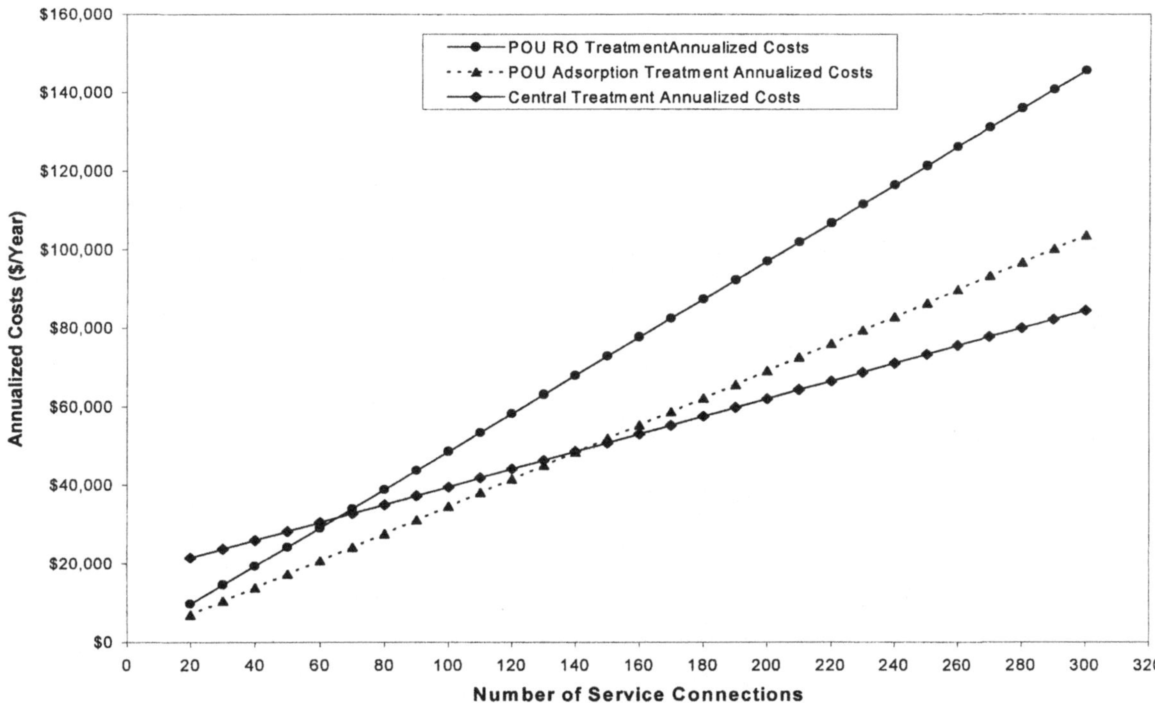

Figure 6.5 Comparison of central and Scenario 1 POU annualized treatment costs (high range values)

Based on a comparison of central treatment and low range POU treatment costs, the breakpoint cost was at approximately 260 connections for POU RO (140 additional connections, when compared to average values) and at 400 connections for POU adsorption (200 additional connections, when compared to average values). However, based on a comparison of central treatment and high range POU treatment costs, the breakpoint cost was at approximately 70 connections for POU RO and at approximately 140 connections for POU adsorption. The cost per month per household (high range) was $40 for POU RO treatment and $29 for POU adsorption treatment. This significant difference in cost (shown in Figures 6.3, 6.4, and 6.5) is due to differences in vendor price quotations, which vary based on lease or purchase option, size of the utility (number of service connections), and accessibility for maintenance activities.

A comparison of central treatment and average POU treatment costs for Scenario 2 was also performed and is presented in Figure 6.6. As shown in the figure, the breakpoint cost was at approximately 160 connections for POU RO (40 additional connections, when compared to average values for Scenario 1) and at 220 connections for POU adsorption (20 additional connections, when compared to average values for Scenario 1).

POE Costs vs. Centralized Treatment Costs

A brief summary of POE treatment costs and a comparison of POE vs. central treatment costs is presented in this section. POE treatment costs were calculated based on the assumption that each household is served by an individual POE, which treats water serving the entire household (for consumption and non-consumption purposes), as opposed to an under-the-sink POU

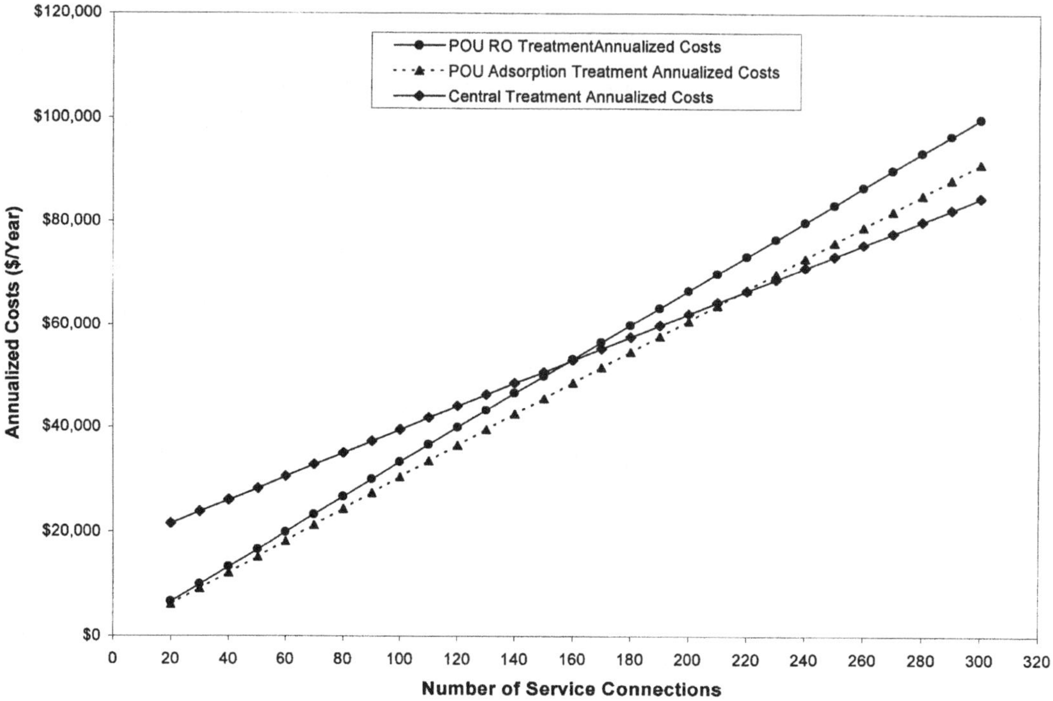

Figure 6.6 Comparison of central and Scenario 2 POU annualized treatment costs (average values)

unit, which treats only water used for consumption purposes. A comparison of POE treatment costs and central treatment costs is presented in Figure 6.7 (for systems where the domestic use represents around 50% of total potable water supply).

The breakpoint analysis, as shown in Figure 6.7 illustrates that POE treatment would not be competitive with centralized treatment. For the smallest size system (20 connections), the annualized cost for the POE option would be close to $41,900, while it is only $21,600 for centralized treatment and between $6,000 and $7,000 for the POU options. The reason for this is that small treatment units have a higher unit cost compared to larger centralized treatment. This difference in costs increases between POE treatment and centralized treatment because there is no economy of scale in POE systems. Moreover, the monitoring, operational and management issues become more and more complex. A cost analysis performed by EPA (Preamble to Final Arsenic Rule, USEPA, 2001) shows significantly lower costs for POE systems. The breakpoint according to EPA occurs at 53 households. However, EPA's costs do not include waste disposal issues, neither do they consider the high labor, operations, and management costs involved in a POE treatment program.

SUMMARY OF POU TREATMENT COST ASSESSMENT

The evaluation of treatment costs for the two different maintenance and monitoring scenarios shows that the cost per household per month for both scenarios is very similar, as summarized in Table 6.6. The tradeoff between the two scenarios, however, is high monitoring accompanied by low maintenance activities for Scenario 1 and low monitoring accompanied by

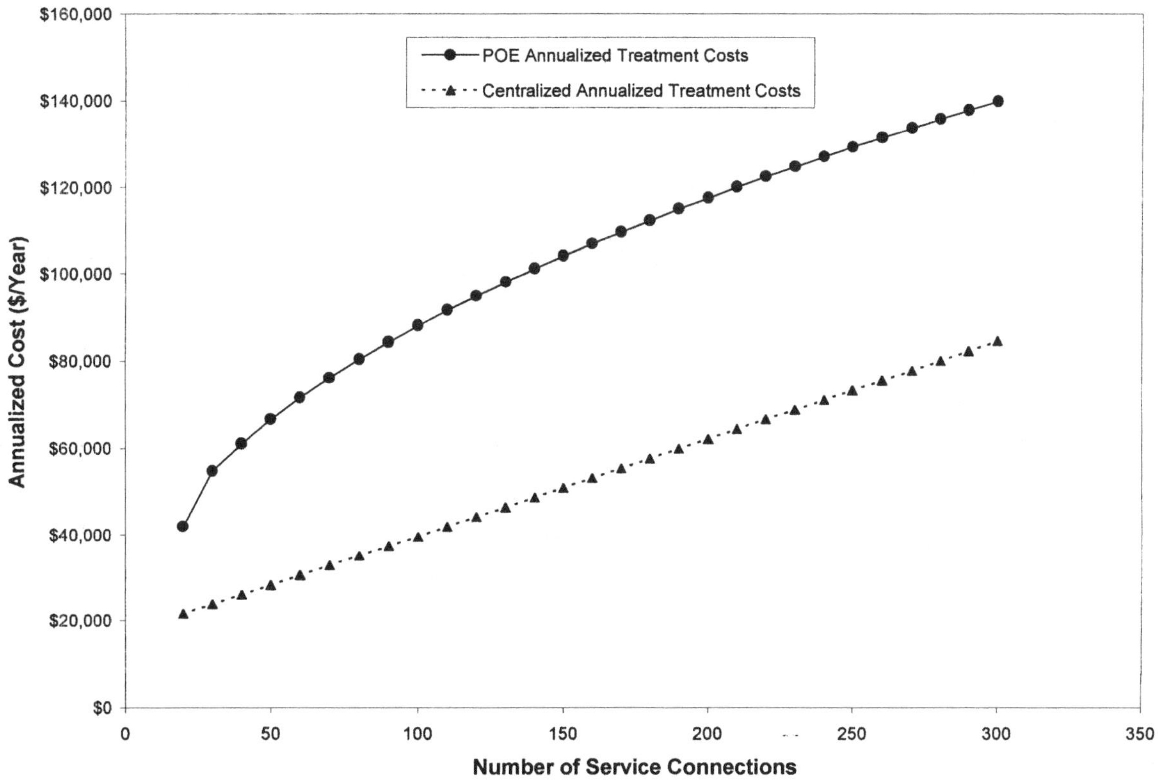

Figure 6.7 Breakpoint analysis for centralized and POE treatment costs

high maintenance activities for Scenario 2. The costs for both scenarios are similar, whether the service is provided by a vendor on a lease basis or if the utility purchases and maintains the units.

A comparison of central treatment costs for throw-away adsorption media and average POU treatment costs shows that POU RO and POU adsorption annualized costs (average values) were significantly lower (35-45%) than central treatment annualized costs for systems serving fewer than 50 connections. As the number of connections increased, the POU costs also increased gradually, and the breakpoint cost was observed at approximately 120 connections for POU RO and at approximately 200 connections for POU adsorption. For systems serving greater than 50 and 60 connections, respectively, annual O&M costs for POU RO and POU adsorption treatment were higher than for central treatment, due to an increase in administrative, monitoring, and replacement costs for membranes, media cartridges, and pre- and post-filters. The capital costs of the POU units, however, were consistently lower than central treatment capital costs. Overall, POU treatment is cost-effective when compared to centralized treatment and is viable for very small systems.

POE treatment costs are generally higher (39-48%) than centralized treatment costs for systems serving 20 to 300 connections. Therefore, POE treatment may not be a feasible option for small water systems, except for cases where a significant portion of the central system's water supply may be provided for irrigation or other non-drinking and non-residential water use purposes, and a savings in treatment of this volume of water can thereby realized by POE treatment.

CHAPTER 7
POU/POE IMPLEMENTATION BARRIERS AND POTENTIAL SOLUTIONS

Members of the research team participated in two stakeholder meetings that were held to obtain input from public officials, utilities, and manufacturers on the factors involved in community use of POU and POE devices for compliance with USEPA regulations, particularly the Arsenic Rule. The first meeting was held at the ASDWA meeting on October 12, 2002. The second meeting was held at the University of Nevada, Reno Arsenic Treatment Workshop (satellite conference) on October 31, 2002. Over 50 attendees were present at each event. A series of questions was posed to the attendees at each meeting to obtain input on the following:

- Monitoring issues (e.g., frequency)
- Liability issues
- Utility responsibilities and administrative requirements
- Changes in technology and NSF certification requirements in the last 20 years (updates to standards, automatic shut-off, etc.)
- State implementation issues
  - Lack of Federal Regulations pertaining specifically to POU/POE devices
- HPC bacteria growth in POU devices
- Safety factor with respect to maintenance frequency (6 months vs. 1 year change-out for cartridges)
- Recalcitrant customers
- Whole house vs. one faucet treatment (POE vs. POU devices)
- Methods for dealing with customers with different usage patterns
- Treatment of non-residential taps (e.g., park fountains, commercial buildings, etc.)
- POU cost feasibility for arsenic treatment
- Disposal issues
- Other issues (e.g., installation)

In addition, the researchers met with state regulators in Arizona, Nevada, and Maine to obtain their input on the issues involved with the use of POE and POU devices for compliance purposes. A summary of the responses from these meetings and the stakeholder meetings and possible solutions to address the key issues are provided in the sections below. The key issues, barriers, and potential solutions are further summarized in Table 7.1. The discussions are focused on the use of POU or POE devices for regulatory compliance with the Arsenic Rule.

REGULATOR AND STATUTORY ACCEPTANCE ISSUES

Implementation of a POU or POE treatment program begins with the state regulatory agency, which, in order to accept POU or POE treatment for compliance and develop policies and guidelines, must first address the issues discussed below. Some regulators and members of the industry feel that these issues should be clarified in the SDWA, to increase confidence in POU/POE programs.

Table 7.1
POU and POE treatment program implementation issues, barriers, and potential solutions

Stakeholder	Issue	Barrier	Potential solution
Regulator	100% household participation	Recalcitrant customers	Issue variances & have recalcitrant customers obtain water at utility; require ordinances making use of water contingent on use of POU or POE unit; require utility to install POE unit
	Ownership of POU or POE Units	State vs. utility ownership; some customers may already have POU or POE unit installed	Utility assumes ownership; customer retains ownership and utility contracts with vendor for maintenance
	Whole House vs. One Faucet Treatment	Approval of only POE; high costs for POE	Utility/customer agreements requiring homeowners to educate household members to use only POU treated tap for drinking & cooking
	Treatment of Taps in Commercial & Public Buildings	High cost of installing POU units for each tap in buildings	Require or recommend one POE unit for each building
	Treatment of Transient Use Taps	High cost and maintenance requirements; transient use not an acute exposure	Require posting of notifications informing public that water from taps may contain arsenic and is for occasional and emergency use only
	Frequency of Monitoring	Frequent monitoring would be costly	Require all units to be monitored once during first year of installation; require or recommend 1/4 or 1/3 of units be sampled on a rotating basis after first year
	Use of Surrogates in Monitoring	Some states may not approve use of surrogates; compliance monitoring is costly	Approve use of arsenic field test kits, conductivity meters and TDS monitors for RO units, etc. to supplement data from reduced routine monitoring
	Maintenance Requirements	Frequent (e.g., every 6 months) maintenance will increase overall cost of treatment program	Require that maintenance frequency be determine by a pilot study or by following manufacturers' instructions
	Administration	Administrative burden and costs may increase	Work with utilities to develop streamlined approval processes; implement an approval or permit fee to offset increased costs

(Continued)

Table 7.1 (Continued)

Stakeholder	Issue	Barrier	Potential solution
	Disposal Requirements	Some spent cartridges (e.g., GIM) may not pass CAWET tests and may have to be disposed as hazardous waste	Regulate disposal of spent media cartridges based on level of arsenic in source water
Utility	100% Household Participation	Recalcitrant customers; retaining participation in program	Include termination of services clause in ordinance; install POE unit; customer education
	Treatment of Taps in Commercial & Public Buildings	High cost of installing POU units for each tap in buildings	Install one POE unit for each building
	Treatment of Transient Use Taps	High cost and maintenance requirements; transient use not an acute exposure	Post notifications informing public that water from taps may contain arsenic and is for occasional and emergency use only
	Installation	Compliance with plumbing & building codes; additional staff training & certification may be needed	Provide additional staff training for certifications; contract with vendor that employs licensed installation personnel
	Monitoring & Maintenance	Requires access to private homes and property	Require customers to sign an access agreement; perform sampling and maintenance during evenings & weekends; frequent communication with customers
Customer	Acceptance of Program and POU or POE Devices	Recalcitrant customers	Ordinance making use of water contingent on installation of POU or POE unit; access agreement with utility; customer education; frequent communication; public outreach
	Preference for POE over POU Device	POE unit and associated maintenance costs are higher than for POU unit	Customer pays difference in equipment, maintenance, and replacement parts costs
	Preference for RO over Adsorption Treatment Unit	RO unit and associated maintenance costs may be higher than for adsorption unit	Customer pays difference in equipment, maintenance, and replacement parts costs
	Use of Untreated Taps for Drinking & Cooking Water	Customers may prefer to use other taps for drinking water or for cooking purposes; flow from POU may be considered too slow	Customer education; utility/customer agreements requiring homeowners to educate household members to use only POU treated tap for drinking & cooking

(Continued)

Table 7.1 (Continued)

Stakeholder	Issue	Barrier	Potential solution
	Refrigerator Connection	Customers may prefer to drink from refrigerator dispenser; copper leaching of supply line	Connect refrigerator to POU/POE treated water supply; replace copper supply line with PVC
Common	HPC Bacterial Growth	HPC bacteria levels in POU/POE effluent may be elevated	Customer education; proper maintenance of POU/POE units; brief flushing of POU/POE taps
Liability	Product Liabilities	Utility could be held liable; resulting lawsuits or settlements could be costly	Utility may need to increase liability insurance coverage; contract with vendor for installation & maintenance of POU/POE units
	Purposeful Tampering	Utility could be held liable; resulting lawsuits or settlements could be costly	Customer education; secure outside POU/POE units; utility may need to increase liability insurance coverage
	Injury, Property Damage Claims	Utility could be held liable; resulting lawsuits or settlements could be costly	Utility may need to increase liability insurance coverage; contract with vendor for installation & maintenance of POU/POE units
	Regulatory Liability	Unfamiliarity with systems; large number of treatment devices increases risk of failure and non-compliance	Central testing of NSF approved POU/POE units prior to installation; perform an engineering evaluation of water system (similar to centralized treatment) to assure performance

100% Household Participation

Most states require 100% household participation for approval. Some states, however, question whether one or a few households should be able to hold an entire community hostage and feel that working towards 100% participation and having ordinances or agreements in place requiring customers to participate is satisfactory. These states recognize that substantial participation is preferable to using central systems only. Issuing variances for non-participating households could be an approach for state regulators to address the issue. Under this scenario, utilities would notify the non-participating households that their drinking water is contaminated with arsenic above the MCL. Utilities could offer these households the opportunity to obtain their water at a tap located at the utility or the opportunity to obtain a POU unit. Customer education regarding the benefits of participation is important to gaining the cooperation of recalcitrant households.

Another approach to address the 100% participation issue is to disconnect service to recalcitrant households. As mentioned in Chapter 2, San Ysidro, N.M. passed an ordinance that made the use of village water contingent upon installation of an RO unit in the home. For customers that do not comply with this ordinance, the village would shut off their water. Providing POE units for recalcitrant households could be another option to bring a system to 100% participation for state approval. This option, however, raises concerns regarding equity of services, as recalcitrant households would be provided the benefit of receiving treated water for every tap in the house, whereas households participating in the POU treatment program would receive supplemental treatment at only one tap.

Ownership of POU Units

For households that already have equivalent or superior POU units installed, questions of ownership and control will need to be resolved. Will the state or utility take over ownership of the POU unit? Many states would probably prefer that the utility assume ownership. Other options could be to let the homeowner keep ownership of the POU unit and have the utility contract with a vendor to maintain the unit, or provide the homeowner with a list of approved vendors to choose from for maintenance. State regulatory agencies could provide utilities with model agreements between homeowners and the utility for maintaining the POU units.

Whole House vs. One Faucet Treatment

Certain states will approve only POE (whole house) systems because they consider health protection substandard when only one faucet is treated. However, the issue of chronic versus acute exposure needs to be considered, particularly for water sources with arsenic contamination. Consumption of an occasional glass of water from a non-POU tap would not constitute an acute exposure risk and may not constitute a chronic risk, depending on the frequency of occurrence. For treatment programs that rely on POU units, customers need to be educated to use only the POU faucet for all drinking and cooking purposes. Idaho, for example, requires that all customers of a water system sign an access agreement (which allows water system personnel to enter their property on a scheduled basis to install and maintain the treatment devices) that includes a statement that the homeowner assumes responsibility for instructing family members or other residents that they should not cook with or ingest water from untreated taps. Cost factors must also be considered. Installation and maintenance of POE units for all households may be cost-prohibitive, particularly for small communities.

Treatment of Taps in Public and Commercial Buildings

Cost is an important factor to consider in regulating treatment of taps in commercial buildings, such as large office complexes, and public buildings (e.g., schools and libraries). POE units may be the most feasible option, rather than a POU unit for every tap, due to the large number of taps in these buildings.

Treatment of Transient Use Taps

Transient use taps include public drinking fountains, which would probably require individual POU units. Depending on the number of public drinking fountains in a community, installation of POU units at each tap could be cost prohibitive. Public fountains generally have transient use patterns, however, and would not constitute a high risk of exposure. Rather than require treatment of each transient tap, regulators could require communities to post notifications near the fountains informing the public that the fountains may contain water with arsenic and are for occasional or emergency use only.

Monitoring Issues

In general, samples should be collected from POE and POU units at the time of scheduled maintenance, but prior to servicing the treatment unit. In theory, there should be no samples that exceed the MCL, because the POU devices must be equipped with a shut-off device or an alarm device to signal treatment failure. Nevertheless, monitoring frequency and the role of surrogates in monitoring are issues that regulators will need to address.

Frequency of Monitoring

Most state regulators appear to prefer providing guidance that ensures compliance and protects public health, rather than regulations that define a required frequency of monitoring. Guidance documents should specify monitoring criteria, which should take into account that a POU or POE system is quite different from a central treatment system. Guidance on frequency of monitoring should also take into consideration the performance capabilities of the POU or POE technology used and the analysis costs, which are high for many parameters, associated with compliance monitoring. In addition, compliance costs must be balanced against maintenance costs. If frequent maintenance is required, increased monitoring may be required, thus driving up costs for both.

The EPA POU/POE guidance document recommends sampling of the effluent from the POU/POE units in each household at least one time during the year in which the units are initially installed. If the treatment units are successful in reducing the contaminants of concern below the MCL, EPA recommends that for chronic contaminants, sampling frequency can be reduced to once every three years, with one-third of all units within a system sampled each year on a rotating basis. For acute contaminants, EPA recommends that sampling be performed more frequently. States may however, choose to require more frequent monitoring. General monitoring requirements in Pennsylvania, for example, include quarterly monitoring on 25% of the installed POE devices, with the locations rotated so that each device is monitored at least once annually.

Appropriate monitoring frequencies will vary between communities depending on raw water quality characteristics, seasonal variations in demand, and exhaustion capacity of RO membranes or adsorption media. The statistical assessment of the results from the in-home POU field testing performed as part of this research project demonstrate that a conservative POU monitoring frequency may be one sample per household during the first year, and one sample per household every 6 months, after the first year. However, a reduced monitoring frequency of sampling only one-fourth of the total number of households per year, accompanied by adsorption media and membrane replacement frequency of once/year once/two years, respectively, may also be acceptable.

Use of Surrogates

Many state regulators feel that the use of surrogate analytes is not acceptable and that compliance monitoring should be required. Some state regulators, however, feel that due to the high costs of compliance monitoring, the use of surrogates will be necessary in testing POU or

POE devices. The role of surrogates in compliance monitoring has been an issue of debate for many years in the water industry. Although further research may be needed to determine how appropriate surrogates are for use in monitoring water quality from POE and POU units, there may be surrogate analytes (e.g., field test strips or gallons of water treated) that regulators can agree to accept in the interim. Many POU devices include a shutoff mechanism that is activated once the unit has treated a certain number of gallons of water (e.g., 500 or 750). In addition, some vendors install RO systems with TDS monitors that compare raw and treated water TDS values.

The EPA guidance document suggests that the quality of the water from the POU or POE units can be spot-checked during routine maintenance visits through the use of commercially available field testing kits, electrical conductivity meters (for RO units), and water hardness testing (for cation exchange units). Arsenic field test kits are commercially available that can detect arsenic in water at levels as low as 2 ppb. These field test kits are available in a wide range of sensitivities (2-160 ppb), and could be used for periodic monitoring by system operators to determine if the POU units are close to reaching breakthough. The results of the water quality spot-checks would be used to augment the POU or POE performance data obtained by routine sampling and laboratory analyses.

Maintenance Requirements

EPA guidance recommends that effective unit capacity (total volume treated below the MCL) of POE and POU units should be determined during pilot testing in order to determine a maintenance schedule. To build in a substantial safety factor, EPA initially recommended performing maintenance every six months. Many state regulators are agreeable to longer maintenance intervals and cartridge change-outs, provided that monitoring is sufficient to demonstrate continued compliance. Most POU and POE units are designed, tested, and certified for treating greater than 1,000 gallons, and many contain an automatic shut-off feature at 500 gallons. Shutting off of the units at 500 gallons constitutes a safety factor greater than the USEPA required safety factor of 20%. Water usage will, of course, vary between customers. However, customer usage patterns would become known to water utilities and the POU or POE vendors, and regular maintenance could be scheduled according to these patterns.

EPA recommends, and many states (e.g., Idaho) require, that water systems provide an operations and maintenance plan which demonstrates that the POU treatment units will be serviced in accordance with the manufacturer's instructions. The Idaho Department of Environmental Quality's POU treatment guidelines (see Appendix B) state that required service intervals will be determined initially by a pilot study or by following manufacturer's recommendations; the maintenance schedule can later be adjusted to reflect actual experience.

Administration Issues

Administrative feasibility, burdens, and costs necessary to approve and implement POU/POE programs are concerns for state regulators. Hiring freezes, lack of funds to hire additional personnel, delays in hiring processes, and the current workload of state regulators could cause delays in approval and implementation of these programs. States should work with utilities and small communities to develop streamlined approval and implementation processes that would require minimal additional personnel for all three. To offset administrative costs, states could consider implementing a POU or POE treatment program permit or approval fee.

Disposal Issues

Most state regulators feel that there would be no significant disposal issues provided that the spent POU or POE cartridges can be disposed of as non-hazardous solid waste. Mn-AA and iron media cartridges are expected to pass the Federal Toxicity characteristic leaching procedure (TCLP) tests, and can be disposed in domestic landfills in most states. TLCP data, from this study confirm that spent aluminum based (e.g., Mn-AA) cartridges can be disposed of in this manner. Spent iron media cartridges (e.g., GFH), however, may contain high enough levels of arsenic to be classified as hazardous solid waste, in California and may have to be disposed to a hazardous waste landfill. Further testing is needed to assess the impacts of varying levels of arsenic in source water on the disposal of spent iron media cartridges, particularly in California.

UTILITY ISSUES

Most utility issues will depend, of course, on the POU and POE approval and implementation process that is adopted by the state. As these processes may vary among states, the resulting utility issues may be different in different states. Nevertheless, there are underlying issues that every utility will face when implementing POU or POE systems for water treatment. Some of these underlying issues are discussed in the sections below.

Household Participation

Utilities will need to develop methods to encourage substantial, perhaps 100% participation in implementation of a POU or POE program. Community systems in Idaho, for example, are required to legally adopt termination of service clauses or ordinances in order to deal with uncooperative customers. If economically feasible, the utility may choose to install POE units at the homes of recalcitrant customers to minimize issues related to accessing the home. Effective public relations and public education programs aimed at convincing customers of the benefits of participating will be essential. Utilities may also have to develop methods to retain customer participation and to deal with customers who "drop out" after the program is implemented. For customers who already have equivalent or better POUs or POEs installed, a utility may need to negotiate an agreement with the customer regarding the maintenance of the unit.

Treatment of Taps in Public and Commercial Buildings

Utilities may be faced with the cost of providing POU or POE units for public (e.g., schools, libraries, day care centers) and commercial buildings (e.g., office complexes). In these buildings, multiple taps may require treatment and costs of POU or POE treatment could be comparable to or exceed the costs for centralized treatment, depending on the number of such buildings in a community. Utilities should carefully evaluate the economic feasibility of a POU or POE treatment program and consider implementing centralized treatment, if necessary.

Treatment of Transient Use Taps

Arsenic treatment costs may be increased if utilities are required to provide POU treatment for public fountains and other transient use taps. POU units at transient taps would need to be frequently monitored for vandalism, even if installed in a protective, secured enclosure. If not required by state regulations to be treated, the utility may need to post notifications near the fountains informing the public that the fountains may contain water with arsenic and are for occasional or emergency use only.

Installation Issues

If utilities choose to handle the installation of POUs or POEs themselves, adequate and appropriate staff training would have to be provided. Compliance with plumbing codes and building ordinances are factors in installation, and utility staff would need proper training and licensing to reduce utility liability. Contracting with licensed plumbers to perform the installation is an option to reduce the amount of utility staff needed and reduce utility liability. Many manufacturer-approved vendors employ installers who have the appropriate training and are licensed plumbers. Utilities that choose to use vendors to install the POU or POE units will need the expertise to develop appropriate contract agreements.

Monitoring and Maintenance Issues

For POU treatment programs, utilities will need to have access to customers' homes in order to perform compliance monitoring and maintenance. For POE systems, utility staff or vendors will need to have access to customers' property and possibly private homes in cases where the POE unit is installed in the garage. Some states (e.g., Idaho) require that water system customers sign an access agreement which allows water system personnel to enter their property on a scheduled basis to install and maintain the treatment devices. In addition, since many customers may not be at home during normal weekday business hours, most sampling and maintenance activities will need to be performed during evenings and weekends. This may make it difficult for a water system to attract and maintain a qualified operator and will add substantially to the cost of any service contract negotiated with a vendor.

Customer relations and communication skills will be important, and utility staff and vendors may need additional training in order to maintain a good relationship with the utility's customers. The water system will need to rely on its customers to help with detecting leaks, noting the status of the alarm device, and promptly communicating any other problems that may be encountered. This requires frequent outreach activities that can be costly and demanding.

CUSTOMER ISSUES

In water systems that utilize central treatment, customers have traditionally had a passive role. For POU or POE treatment programs to be successfully implemented, the customers must be actively involved, and the water system needs to partner with its customers to address the issues discussed below.

Customer Acceptance

Customer acceptance of POU and POE devices will obviously affect implementation of a POU or POE program. Some customers may not be familiar with the technologies and may be reluctant to have a "gizmo" in their house. Many POU units take up space under the kitchen sink or on the countertop near the sink, and customers may be unwilling to give up that space. Some customers will consider regular maintenance inconvenient and it may be difficult for service personnel to gain access to those homes. In addition, for POU programs, customers may be concerned that only one tap in the home is treated by a POU unit.

Many of these customer acceptance issues can be addressed by effective communication and education by the utility and, in some cases, by the manufacturers and vendors. Customers should be educated on the capabilities and maintenance requirements of the technologies. Utilities and vendors could team up to provide public demonstrations of the POU or POE units. Customer education should also include information on the cost of health versus inconvenience trade-offs and chronic versus acute exposure risks. Effective communication with customers will aid in establishing relationships between the utility and vendor service staff and the customers. This could, in turn, help service staff gain access to the homes to perform the necessary maintenance. Other approaches to promote customer acceptance of POU and POE treatment programs could include the following:

- Customer education using techniques such as consumer confidence reports, advertising, and public service announcements
- Promote third-party certification of POU and POE products
- Publicize POU and POE certification requirements
- Develop community outreach programs
- Seek out health professionals to assist with public education
- Publicize demonstrations and describe other communities where POU/POE programs have worked
- Provide U.S. EPA certification for POU/POE programs

POU vs. POE Devices

Because of cost and waste disposal issues, POE treatment may not be suitable for many small water systems. However, some customers may decide that they would like supplemental treatment for every tap in the house and that they prefer to have a POE unit rather than a POU unit. In addition, customers who choose not to participate in a POU treatment program may be willing to accept a POE unit in order to minimize access by utility and vendor personnel to their home and property. Utilities with POU programs in place could deal with these issues by offering these customers the chance to upgrade, if the customers are willing to pay the difference between a POU and a POE unit. The price difference would need to include not only the unit but also replacement media/parts, installation, and maintenance costs.

RO vs. Adsorption Treatment

Some customers may prefer to use a RO POU device to gain additional water quality benefits. For these customers, a water system that has chosen to implement a POU treatment program using adsorption media could install RO units, providing that the customers are willing to pay the differences in costs for the RO POU unit, installation, replacement parts (e.g., post and pre-filters and membranes), and maintenance.

Use of Untreated Taps for Cooking and Consumption Purposes

For various reasons, customers may drink from untreated taps or use untreated water for cooking purposes. As seen in the in-home POU field testing program, several homes with low water usage during the study indicated that members of the household preferred drinking the water dispensed from the refrigerator rather than the POU. In addition, several homeowners used regular tap water for cooking purposes because the flow rate from the POU tap was low and it took too much time to fill a pot. Effective and frequent outreach activities are important to educate customers concerning the health risks associated with the local contaminant(s) of concern and the importance of continually using POU treated water for drinking and cooking. Water systems may elect to place the burden of responsibility for instructing family members or other residents that they should not cook with or ingest water from untreated taps on the homeowner. Idaho, for example, requires that water systems enter into legal agreements (with customers) that include a statement that the homeowner assumes the responsibility for educating family members or other residents on this issue.

Refrigerator Connection

As shown in the results from the in-home testing program, some customers prefer to drink water from the dispenser in the refrigerator door. Many customers with modern refrigerators may prefer that it is connected to the POU or POE treated water supply. Commonly, the supply line from the refrigerator, as supplied by the manufacturer, is copper. As copper leaching may be an issue, particularly for RO treated water, utility or vendor installation personnel should replace the supply line with plastic tubing to minimize leaching.

COMMON ISSUES

HPC Bacteria Levels in POU/POE Effluent

Historically, heterotrophic plate count (HPC) bacteria levels have been used as a secondary measure of disinfection efficiency. Under EPA's surface water treatment rule, water systems using surface water or ground water under direct influence of surface water must consider HPC levels in their overall compliance criteria, with a target level of 500 CFU/mL. In most cases the water entering a POU or POE device will meet this requirement; however, the levels of HPC in the effluent from POU and POE units are commonly higher than the levels found in the water supplying the units. This is not surprising, since POU and POE devices reduce or remove the disinfectant residual. Further, GAC units may provide a carbon food source for bacterial growth and increased surface area for bacterial attachment. For waters with a chlorine residual less than 0.1 mg/L, HPC becomes a concern to utilities and regulators when the level is greater than 500 CFU/mL.

The value of performing HPC analyses is still under debate in the drinking water community. International experts at an international World Health Organization (WHO) workshop concluded that increased HPC levels in effluent from domestic water treatment devices do not indicate a health risk if the influent water meets other microbial standards (WHO, 2002). The committee suggested that appropriate maintenance of the devices, according to the manufacturer's recommendations, is required for aesthetic reasons. Although many experts in the scientific community concur with the WHO committee's conclusions regarding the public health risk associated with elevated levels of HPC bacteria, when considering implementation of a POU or POE program, regulators, utilities, and customers are unlikely to accept water quality that is different than the quality of water currently being served to the customer.

It is likely that, in most states that allow (or plan or propose to allow) POU or POE devices for compliance, the utility will be responsible for operation and maintenance of the POU/POE devices, as shown in the results of the state regulatory survey discussed in Chapter 3. With this responsibility, utilities are not likely to find large increases in HPC from POU or POE devices acceptable. The "less than 500 colonies/mL mentality" will not change until regulators develop specific HPC related standards for POU/POE effluent or remove responsibility for maintaining the devices from the utilities. Many customers, even if they are aware of the lack of evidence of health effects, likelihood of bacterial growth, and POU/POE maintenance requirements, may not understand and be willing to accept large increases of bacteria in their drinking water. Customers will expect utilities to be able to offer solutions to this situation.

Brief flushing of the POU/POE taps may be sufficient to lower HPC levels. Data from in-home POU testing for the AwwaRF project *Conventional and Unconventional Approaches to Water Service Provision* (Raucher et al. 2004) demonstrates that the HPC and coliform counts in the POU effluent are reduced after a brief flush (five to ten seconds) of the tap. Some customers are already aware of the benefits of flushing and practice it on taps and drinking fountains supplied by centrally treated water. Although flushing of POU or POE taps may be seen as wasteful, the effluent could be collected and used for a number of purposes (e.g., plant watering, beverages that require boiled water, etc.). Customers could be offered this solution and educated in conservation strategies. Another possible approach could be for manufacturers to design POUs and POEs with "filter to waste" capabilities. When the POU/POE tap is first turned on, effluent from the POU or POE device could be routed to a drain line for disposal.

Regardless of the solution, utilities will need to work with manufacturers and vendors in order to educate customers on proper maintenance of the POU and POE units.

LIABILITY ISSUES

Product liabilities exist with both POU and POE units. With POU units, there would be exposure paths for untreated supply, since it is likely that only one tap in a household would contain a POU unit. Liability for purposeful tampering may also exist, regardless of the security involved. In addition, there is the issue of equipment failure, claims of injury, property damage claims, as well as installation, maintenance, and other such liabilities. As shown in the in-home POU testing performed at the Silverbell Water System in Tucson, Ariz., several of the POU units leaked and caused damage to the cabinets under the customers' kitchen sinks. In such cases, the utility may need to reimburse the customers for the damage. Many of these liabilities are probably insurable. Utilities may also be able to reduce some of these liabilities by contracting with manufacturer approved vendors for the installation and maintenance of the POU and POE units.

State regulators may also be faced with liability issues. There may be a regulatory liability that relates to the unfamiliarity with the systems until completely accepted by state regulation. The liability of approving POU/POE systems by state agencies and the resultant liability to the general public and special interest groups is also a concern to regulators. With a large number of treatment devices, the risk of failure, and possibly noncompliance, is increased. The potential solution to this problem would be to rely on central testing of ANSI-accredited third party tested and certified POU/POE units prior to installation. Additionally, an engineering evaluation should be performed (similar to centralized treatment), to assure system reliability and performance.

SUMMARY OF IMPLEMENTATION BARRIERS AND POSSIBLE SOLUTIONS FOR POU/POE TREATMENT PROGRAMS

Development and implementation of POU or POE treatment programs is complex and involves regulatory, utility, customer, liability, and other issues. Regulatory agencies are faced with addressing issues such as treatment of taps in public and commercial buildings, 100% household participation, health risks associated with use of non-treated taps, treatment of transient use taps (e.g., park fountains), required frequency of POU/POE monitoring and maintenance, disposal of wastes, and increased administrative burden and costs. Some of these issues may be addressed by requiring ordinances that make use of community water contingent on installation of a POU or POE unit, or by requiring agreements between water systems and customers to allow access to homes to maintain the POU or POE units. Utilities or water systems must address issues including household participation and retention in a program, providing treatment to taps in public and commercial buildings, treatment of transient use taps (e.g., park fountains), and gaining access to homes for installation, monitoring, and maintenance purposes. Customer related issues play an important role in the development of a successful POU or POE treatment program. These issues include customer acceptance, preference of a POE over POU device, and, in the case of a POU treatment program, use of untreated taps for consumption and cooking purposes. Liability issues such as product liability and liability associated with entering homes for installation, monitoring, and maintenance purposes may be addressed through adequate liability insurance and contracting services with licensed vendors. Customer education and communication are key to overcoming the majority of barriers to implementation of a successful POU/POE treatment program. Effective outreach will help in developing a partnership between regulators, water systems, and customers.

CHAPTER 8
MANAGEMENT AND IMPLEMENTATION GUIDELINES FOR UTILITIES

This chapter synthesizes information from the previous chapters in order to develop guidelines that will be helpful to water utilities for implementing and managing POU or POE treatment programs. Implementation and management of a POU or POE treatment program involves two major steps. The first step is to determine whether the water system is a viable candidate for a POU or POE treatment program. To determine this, a water system needs to be able to positively answer several questions. The second step is implementation of the program, including communication with regulators, field testing of a POU or POE technology to demonstrate performance under site-specific field conditions, and development of a management plan. This chapter describes the questions that need to be considered in the first step, and provides guidelines for the specific steps involved in the implementation and management of a POU or POE treatment program. While the guidelines are focused on implementation of POU or POE devices for arsenic treatment to comply with the Arsenic Rule, many of the guidelines may be applicable to other contaminants and regulations as well.

IS THE WATER SYSTEM A CANDIDATE FOR A POU OR POE TREATMENT PROGRAM?

A water system that is interested in implementing a POU or POE treatment program must first determine whether the system is a viable candidate for such a treatment program. In order to determine this, the water system should be able to positively answer the majority (i.e., at least six out of eight) of the following questions.

Does the State Approve and Support Use of POU or POE Treatment Devices for SDWA and Arsenic Rule Compliance?

If the State does not approve POU or POE treatment for the Arsenic Rule and other SDWA compliance purposes, the water system will not be able to go any further. The water system should contact the appropriate State regulatory agency (e.g., Department of Health Services or Department of Environmental Quality) to determine the State's rules and policies and obtain any available guidance. If the State does approve the use of POU or POE units, the water system should start a dialogue with the State agency regarding the proposed treatment program.

Is Arsenic the Only Contaminant of Concern?

The water system should characterize the quality of the source water and determine if any contaminants other than arsenic are present at levels exceeding their respective MCLs. If the water system has water quality issues and treatment needs in addition to arsenic, this may determine whether POU or POE treatment is even feasible. In some cases, POU or POE technologies may not be able to effectively treat multiple contaminants. High levels of iron and manganese in the source water may foul both RO and adsorption units. Some contaminants (e.g., radium, uranium, chromium, and fluoride) may not be effectively removed by adsorption devices, while they are effectively removed by RO. If other contaminants also need to be removed, RO technology is preferred.

For source waters with radon or VOCs, USEPA guidance states that POU treatment should not be used. In addition, treatment of microbial contaminants using POU devices is not allowed by SDWA. Depending on state regulations, POE treatment may be an option for radon, VOCs, or microbial contaminants, however cost and waste disposal issues associated with POE devices must be considered.

Is the Source Water Suitable for POU or POE Arsenic Treatment?

The presence of other water quality parameters in the source water that will interfere with arsenic treatment will determine what type of POU or POE device (i.e., adsorption or membrane) should be considered. The primary parameters of concern for POU adsorption devices are silica, fluoride, iron, manganese, and sulfides. For POU membrane systems, the primary parameters of concern are iron, manganese, sulfates, hardness, barium, silica and NOM. For a POU adsorption system, where only a small fraction of the flow is treated, a higher level of fouling can be tolerated and the expectation for system performance is around 1,000 BVs. However, if the interfering parameters are present in extremely high levels, the POU system may not perform adequately to treat even 1,000 BVs.

POE systems may be more similar to central treatment systems where arsenic removal is affected by the presence of water quality parameters such as silica, phosphorus, pH, fluoride, sulfate, chloride, TDS, sulfides, iron, and manganese. Parameters that may lead to fouling of membranes include iron, manganese, calcium, magnesium, barium, silica, and NOM. Water quality parameters that affect arsenic removal efficiency are pH and alkalinity. In adsorption systems, parameters that compete with arsenic for removal include silica, vanadium, phosphorus, fluoride, sulfides, sulfate, chloride, and TDS. For example, silica levels greater than 50 mg/L and fluoride levels greater than 2 mg/L affect adsorption using alumina based sorbents, and pH values greater than 8.5 and phosphorus levels greater than 0.5 mg/L affect iron based adsorption units. A generic list of maximum acceptable levels of these interfering parameters is shown in Table 8.1 for adsorption and RO systems (based on the findings of this study). These maximum acceptable levels are site specific and should be verified by field testing.

No matter what type of POU or POE technology a water system chooses to use, on-site field testing should be performed prior to implementation in order to verify the performance of the technology and determine the impacts of any interfering parameters.

Is the Size of the Water System Favorable for POU or POE Treatment?

The size of the water system will determine whether implementation of a POU or POE treatment program is economically feasible. As shown in Figure 6.3 in Chapter 6, the break-even point for annualized costs for POU treatment when compared to costs for centralized treatment is at approximately 120 connections for POU RO and at approximately 200 connections for POU adsorption. Significant cost savings may not be realized for POU treatment unless the number of connections is at least 20% lower than the break-even point. Thus, it may not be economically prudent to consider POU treatment if the water system contains much over 200 connections, This number will vary depending on local market conditions as shown in the sensitivity analyses presented in Chapter 6.

Table 8.1
Interfering parameters and maximum acceptable levels for arsenic treatment by adsorption and RO systems

Parameter	Maximum levels for Mn-AA, Fe-AA or GIM adsorption	Maximum levels for RO membranes
pH	8.5	NA*
Iron	0.3 mg/L	0.3 mg/L
Manganese	0.05 mg/L	0.05 mg/L
Silica	50 mg/L	50 mg/L
Fluoride	2 mg/L	NA
Sulfides	0.2 mg/L	NA
Phosphorus	0.5 mg/L for Mn-AA & Fe-AA; 0.2 mg/L for GIM	NA
Calcium	NA	150 mg/L†
Magnesium	NA	150 mg/L†
Barium	NA	2 mg/L†
Chloride	250 mg/L	250 mg/L
TDS	750 mg/L	1000 mg/L
Vanadium	0.02 mg/L for GIM	NA
Nitrate	NA	NA
NOM	NA	site specific
Sulfate	200 mg/L	500 mg/L

* not applicable
† general guidelines, but will vary according to site-specific water quality

POE treatment costs may be considerably higher than either centralized treatment or POU treatment, unless savings by avoiding treatment for a significant volume of non-residential and non-drinking water supply can be realized by using POE versus central treatment.

Other limitations to using POE treatment include space and operational constraints. A POE AA treatment unit would be similar to a pilot AA unit, which requires a minimum footprint of three square feet. A 5-gpm unit would be required for a typical home because the unit would have be designed for peak flows (to avoid storage tanks). Also, regular maintenance such as back-washing of the media would have to be performed monthly; this would pose logistical problems for a water system, even in a community of only 20 homes. As with POU units, monitoring at each home would be required, and access issues would need to be addressed.

However, in some situations, such as small transient systems (e.g., roadside stores or restaurants), a POE system may make better operational and economic sense than POU devices. Another example is an apartment complex, where there are a number of families and persons living in a single building. However, apartment buildings may be rare in small systems and rural communities where POU or POE systems are being considered.

The water system will need to consider the above information and determine whether POU, POE, or centralized treatment would be the most economical option for the number of connections present. Anticipated growth of the community in the near future may also need to be factored into the decision.

Do POU Treatment and the Associated Cost Savings Represent the Only Practical Solution for Arsenic Compliance?

The water system should consider whether the cost savings are the primary driver for implementing a POU treatment program. However, cost savings must be balanced against operational and maintenance requirements and public perception issues. For most small water systems, central treatment would make water rates unaffordable for customers; however, some water systems may prefer central treatment to avoid the issues associated with accessing customers' homes for monitoring and maintenance of the treatment devices. In addition, in some communities, customers may not be willing to accept treatment of only a portion of the flow in their homes, and may be willing to pay higher rates for central treatment. Due to economic considerations, POU treatment may be the only feasible option for the system to comply with the Arsenic Rule. If this is the case, this is strong support for a POU type approach.

Will the Water System's Customers Support and Participate in a POU or POE Treatment Program?

Before implementing a POU or POE treatment program, the water system will need to survey its customers to find out whether they would support this type of program and whether they are willing to participate to ensure that the program is successful. Customers must be provided adequate information regarding the purpose of the program, including the health risks of the contaminants for which treatment is needed, and the level of effort that will be required of them. The water system should also provide information concerning the costs of the POU or POE program in comparison to central treatment. Community buy-in and a partnership with the customers will need to be forged before a water system can implement the program. Water systems that are unable to gain customer support may need to consider a centralized treatment option.

Is the Water System Prepared to Manage the POU or POE Treatment Program?

The water system should consider whether it has the resources to manage a POU or POE treatment program. Adequate staffing will be needed for the time required to develop the overall management plan, obtain regulatory approval of the program, perform public outreach activities, negotiate and manage vendor agreements, manage or perform installation of the POU or POE devices, contact customers to set up appointments for sampling and maintenance, and complete the reporting requirements. As mentioned in the USEPA guidance, some water systems have found it helpful to employ a person part-time to develop schedules for installation, maintenance, and sampling, and set up and confirm all appointments with customers and record keeping. For the in-home field testing at the Stagecoach General Improvement District, utility staff estimated that eight to nine hours per month was required for sampling and maintenance related activities (e.g., sample preparation, paperwork, sample delivery to laboratory, filter replacement, etc.) for the 19 homes that participated. An estimated three hours per month was spent by Stagecoach utility staff on customer outreach and to address questions and complaints. Staffing requirements will, of course, vary depending on the size and location of the system and type of program (e.g., POU vs. POE) and type of treatment devices (e.g., RO vs. adsorption) implemented. The number of hours spent by utility staff in Stagecoach, Nev. represents a minimum, as the utility staff did not develop formal management or sampling plans, set up the agreements with the vendors, or perform the reporting that typically would be required by State regulators. Staffing requirements and resources for hiring additional staff, if necessary, will need to be considered by the water system and compared with the requirements demanded by other treatment options.

Are Vendors Available in the Area of the Water System to Support the Program?

The water system will need to investigate whether there are any POU or POE vendors within a reasonable distance that are available to support a treatment program. Although POU and POE devices can be purchased via mail order or the Internet and delivered to the water system, for purposes of installation, maintenance, and training, it is helpful if a vendor is located geographically close to the water system. Many vendors charge extra for mileage and travel time, and water systems that are remote will need to factor in these costs when determining the economic feasibility of a POU or POE treatment program. Travel costs for utility staff to attend training sessions at vendor or manufacturer sites will also need to be considered. It is recommended that the utility contact the potential vendors to make sure they can support such a program.

IMPLEMENTATION STEPS FOR A POU OR POE TREATMENT PROGRAM

Once a water system decides that it is a candidate for a POU or POE treatment program, certain steps need to be followed in order to implement the program. The following sections describe these basic steps. Specific details required in each step will vary depending on the State's regulations and policies; therefore the actions involved in the steps are generalized to apply to the majority of water systems.

Communicate With State Regulatory Agency

The first step that a water system should take in implementing a POU or POE treatment program is to communicate with the appropriate State regulatory agency (e.g., Department of Health Services or Department of Environmental Protection). As mentioned above, contacting the agency to determine the State's rules and policies provides an ideal opportunity for the water system to start a dialog with the State agency regarding the proposed treatment program. It is important to gain the State's support of the treatment program as early as possible in the implementation process. Specific guidance regarding the development of a management plan, including sampling and maintenance plans and reporting requirements, should be obtained to facilitate completion of subsequent steps in the process.

Conduct Pilot Testing of POU or POE Treatment Technology Under Site-Specific Field Conditions

Based on the questions described above, the water system should determine the most appropriate type of POU or POE treatment technology for the water source. This technology should be tested under site-specific field conditions in a pilot study to verify the technology's performance and determine the number of gallons that can be treated before reaching arsenic breakthrough. The scope and duration of the pilot study will be determined by such factors as the characteristics of the raw water and manufacturer's ratings of the treatment device. The results of the pilot study will provide data on any site specific problems with component fouling or microbial growth, determine maintenance intervals, and help in specifying and calibrating alarm or shut-off systems.

Develop Ordinances and Agreements That Will Allow the Water System Access to Private Residences or Property

For legal access to private residences or property in order to install, monitor, and maintain POU or POE devices, water districts and municipalities may need to develop and have ordinances passed in their city or district. Private water systems may need to modify their existing contract with the community in order to gain this access. In addition to allowing access to homes or private property, the ordinances or contract amendments could provide a way to deal with recalcitrant customers. As in San Ysidro, N.M., the ordinance could stipulate that water use is contingent upon installation of the POU or POE unit in the home; noncompliance with the ordinance would result in shutting off the water to the home. The water system should investigate the relevant municipal or county codes or private contract to determine whether ordinances or contract amendments will be required. As it may require significant time to pass an ordinance or complete contract amendments, these should be developed and presented to the community or district early in the process.

Some states require that water systems have every customer commit to and sign an access agreement which allows water system personnel to enter their property on a scheduled basis to install and maintain the treatment devices. In some states (e.g., Idaho), this agreement must include a statement that the homeowner assumes responsibility for instructing family members or other residents that they should not cook with or ingest water from untreated taps (IDEQ, 2000).

Develop and Submit Management Plan to State Regulatory Agency

The water system will need to develop a management plan for the POU or POE treatment program that includes installation and reporting protocols, and training, monitoring, operations and maintenance, public outreach, and customer communication plans. The installation protocol and operations and maintenance plan should consider whether personnel to perform these tasks will be required to be certified as licensed plumbers and/or electricians. The water system should contact local licensing and permitting authorities to verify the requirements of all local codes. The management plan should also include data from the characterization of the source water and data from the pilot testing of the POU or POE technology under site-specific field conditions. In addition, copies of any ordinance passed and customer agreement developed to allow access to install, monitor, and maintain the treatment units should be included. A comprehensive management plan will demonstrate that the water system has the necessary resources to manage a POU or POE treatment program, and will aid in gaining the support of the State regulatory agency. Once approval from the State is obtained, the water system can continue with the implementation of the treatment program.

Perform Public Outreach

Based on the previously developed public outreach plan, the water system should perform public outreach activities to inform customers of the POU or POE treatment program and their role in the program. The water system should provide its customers with educational materials (e.g., in water bills or via flyers) concerning the health risks associated with the contaminant(s) for which treatment is proposed. Customer education should also include information on chronic versus acute exposure risks and the costs of health versus inconvenience trade-offs. Other approaches to customer education to promote the POU or POE treatment program could include the following:

- Consumer confidence reports, advertising, and public service announcements
- Publicize POU and POE certification requirements
- Develop community outreach programs
- Seek out health professionals to assist with public education
- Public demonstrations of POU or POE devices to demonstrate the capability of the technology and maintenance requirements. Utilities and vendors could team up to provide these public demonstrations
- Provide information on other communities where POU/POE programs have worked.

The water system will also need to provide any informational materials to homeowner's associations, realtors, and other groups (e.g., churches, service organizations) within the community to ensure that the information is disclosed to all incoming residents who were not present when the treatment program was first implemented.

Effective public outreach and communication with customers will aid in establishing good relationships between the utility and vendor service staff and the customers. This will allow the water system to gain the confidence and support of its customers and the ability to meet the requirements of the treatment program to comply with the Arsenic Rule and achieve other water quality objectives.

Negotiate Service and Access Agreements

After the water system has located a suitable vendor, or several vendors, that can supply the POU or POE treatment units and provide the desired services, a service agreement will need to be negotiated. Water systems may find it helpful to evaluate the experiences of other systems. The USEPA guidance document (USEPA, 2002) provides information on costs associated with implementing a POU or POE treatment strategy and potential funding sources, and summarizes the experiences of more than 20 communities that have installed POU and POE devices to treat a wide variety of contaminants.

Water systems should obtain bids and references from several vendors, if possible. The water system will need to determine if it is more economically feasible to lease or purchase the treatment units. Vendors may be willing to offer a volume discount and negotiate a discounted service rate if the term of the service agreement is longer (e.g., 3-5 years). When obtaining bids from vendors or manufacturers, water systems should request information on product warranties and the availability and cost of replacement parts.

As a follow-up to the initial public outreach activities, water systems may need to have their customers sign an agreement that allows water system or vendor personnel access to their property on a scheduled basis to install, monitor, and maintain the treatment devices. These agreements should be in place prior to installation of the treatment units, and customers should be informed of the consequences of non-compliance with the agreement.

Participate in Training Activities

As a minimum, water system personnel should be trained in the monitoring and emergency maintenance of the POU or POU devices. For those systems that choose to perform installation and maintenance themselves (i.e., in-house), staff will require additional training and may need to be certified as licensed plumbers and/or electricians, depending on local codes. Water systems may be able to obtain basic training in installation and maintenance of the units from vendors or manufacturers, and water systems should investigate the possibility of including training in purchase, lease, or service contracts.

Perform Follow-Up Evaluations of the POU or POE Treatment Program

A follow-up evaluation of the treatment program should be performed every three to five years to verify that the program is still the most feasible option for complying with the Arsenic Rule. Any growth of the community should be factored into the evaluation. The water system should also investigate new available POU or POE technologies that may be more efficient for arsenic removal or more cost-effective. Lease and service agreements with vendors should be evaluated to rate the performance of the vendor. The training, monitoring, and operations and maintenance plans should also be reviewed to ensure that all requirements are being met. In addition, the water system should evaluate its public outreach program and communicate with customers to determine whether customers are satisfied with the treatment program.

SUMMARY OF MANAGEMENT AND IMPLEMENTATION GUIDELINES

Prior to implementing a POU or POE treatment program, a water system must determine if it is a candidate for such a program. To be a candidate, a water system should be able to answer the majority of the following questions in a positive manner:

- Does the state approve and support use of POU or POE treatment devices for SDWA and Arsenic Rule compliance?
- Is arsenic the only contaminant of concern?
- Is the source water suitable for POU or POE arsenic treatment?
- Is the size of the water system favorable for POU or POE treatment?
- Do POU treatment and the associated cost savings represent the only practical solution for arsenic compliance?
- Will the water system's customers support and participate in a POU or POE treatment program?

- Is the water system prepared to manage the POU or POE treatment program?
- Are vendors available in the area of the water system to support the program?

After a water system has determined that it is a candidate for a POU or POE treatment program, the steps listed below should be followed for implementation of the program.

- Communicate with the State regulatory agency to gain support for the program and obtain guidance.
- Perform pilot testing of the POU or POE treatment technology under site-specific field conditions to verify on-site performance and identify potential problems.
- Develop ordinances and agreements that will allow the water system access to private residences or property for installation, monitoring, and maintenance of the treatment units.
- Develop and submit a comprehensive management plan to the State regulatory agency.
- Perform public outreach activities to educate customers regarding the purpose and importance of the program and their role in the program.
- Negotiate service agreements with vendors and access agreements with customers.
- Participate in training activities to ensure that water system personnel are adequately prepared and certified, if necessary.
- Perform follow-up evaluations of the POU or POE treatment program to verify that it is still the most feasible option for compliance with the Arsenic Rule and other water quality objectives.

The estimated time frame to implement a POU/POE treatment program is one year, considering field testing, customer interaction, state approval, vendor procurement, and installation. Utilities should plan accordingly is POU/POE is selected as a compliance tool.

A flowchart diagram representing a summary of the POU/POE management and implementation guidelines is presented in Figure 8.1.

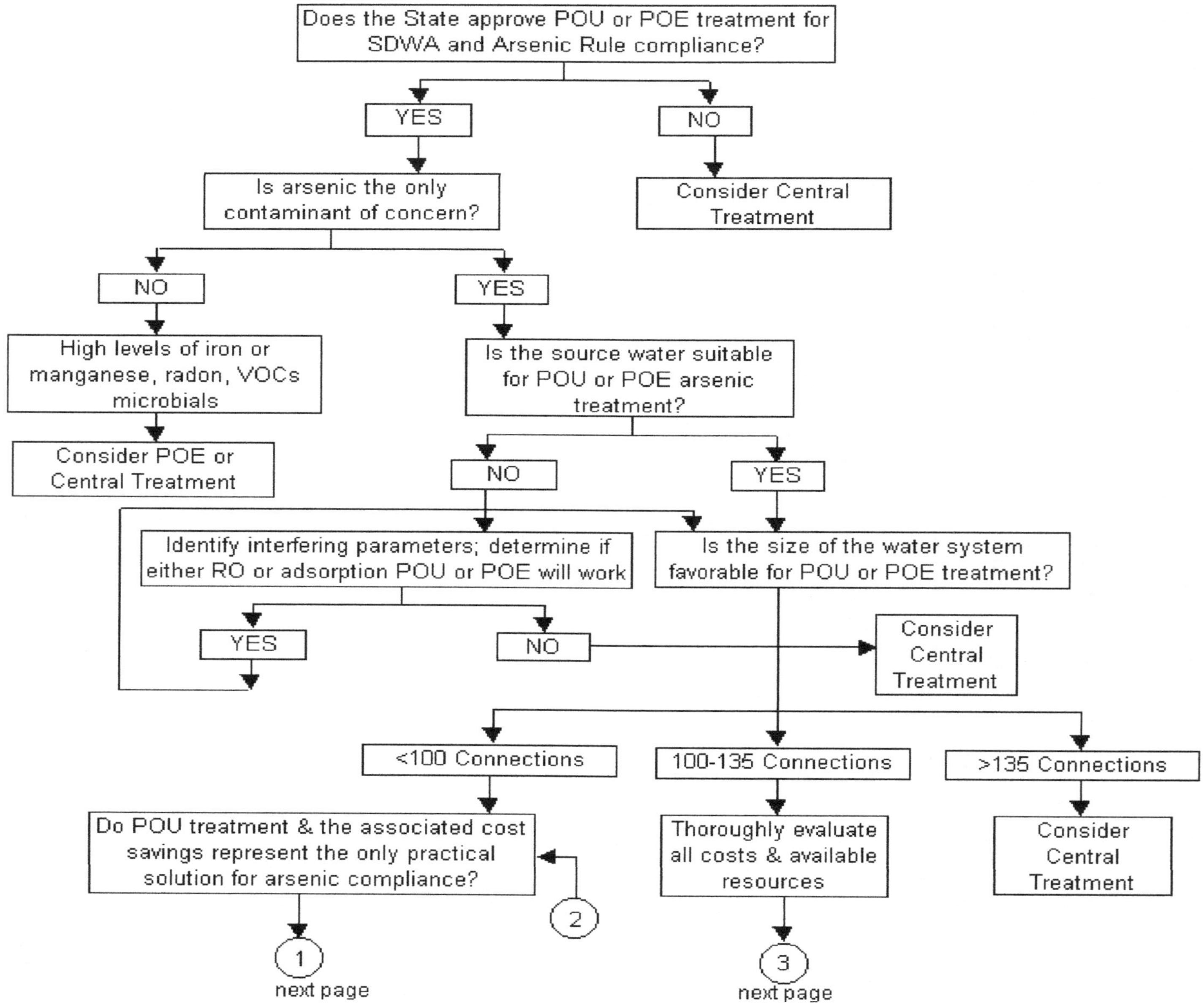

Figure 8.1 Summary of POU/POE management and implementation guidelines

(Continued)

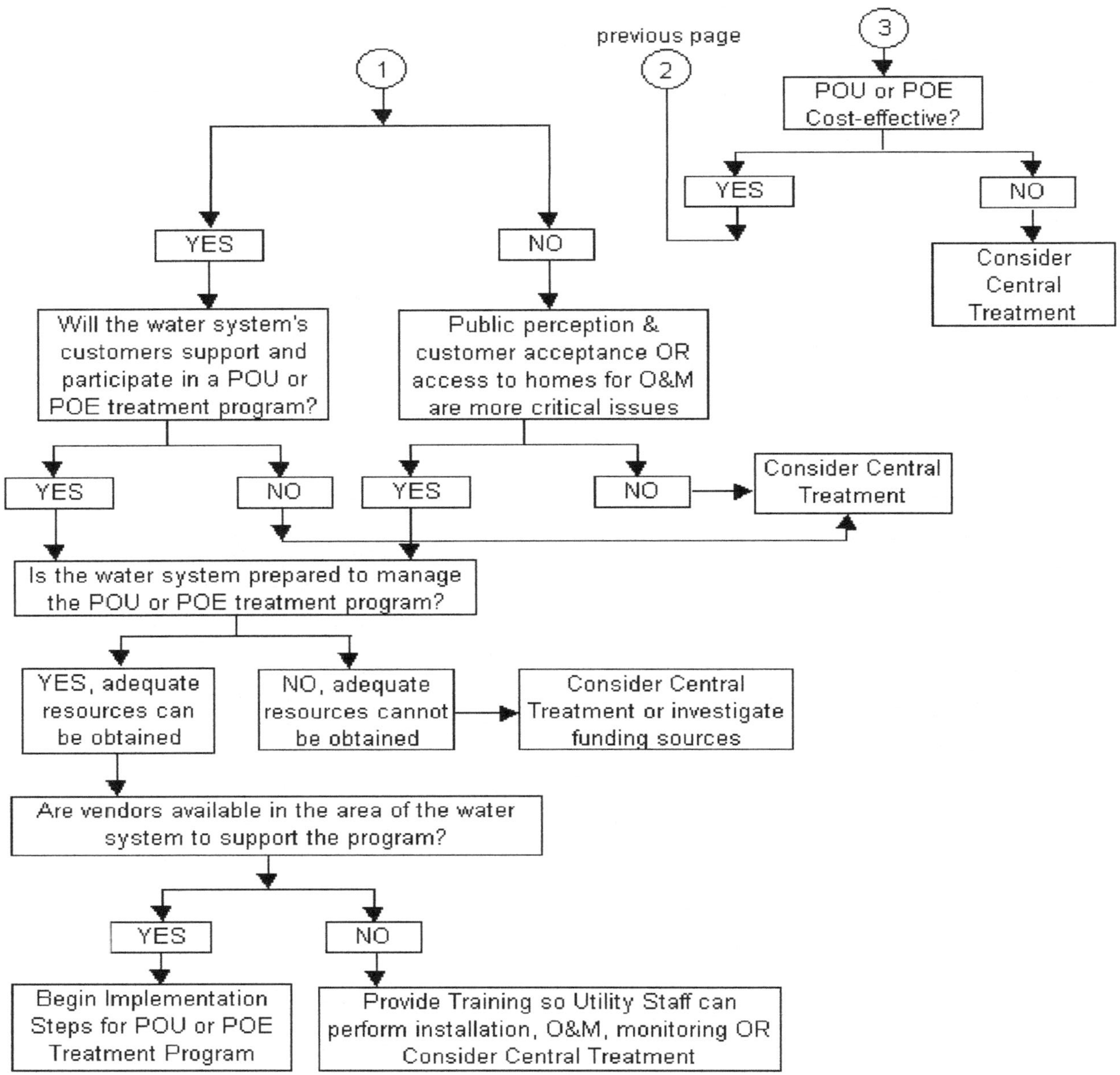

Figure 8.2 (Continued)

CHAPTER 9
CONCLUSIONS AND RECOMMENDATIONS

This report presents the findings of a research study that was conducted to determine the feasibility of implementing POU or POE treatment for arsenic compliance. The major objectives of the study were to develop industry-wide recommendations for use of POU and POE systems as an alternative to centralized treatment, considering such factors as process reliability, public perception, liability, and regulatory acceptance, to assess water quality criteria that may limit the performance of POU and POE systems, and determine at what system sizes POU and POE treatment systems are cost effective, considering management, maintenance, and monitoring expenses. A field screening evaluation was conducted at six different participating utility sites to assess the performance of several POU and POE technologies. The performance of POU units was further evaluated in testing at homes in communities served by three different water systems.

SUMMARY OF RESEARCH ACTIVITIES AND CONCLUSIONS

The findings from the review of previous studies, regulatory review, field screening evaluation, and in-home POU testing formed the basis for developing guidelines for implementing a POU or POE treatment program for arsenic compliance. The major conclusions from the study are summarized below:

- The monitoring results from the in-home POU testing indicated that RO and adsorption POU devices perform well and reliably remove arsenic for periods over one year.
- If the appropriate technologies are installed by trained technicians, POU and POE devices can effectively remove contaminants, including arsenic, and improve the taste and odor of the water.
- Successful performance of POU systems is strongly dependent on an active operation and maintenance program, performance testing of the systems, and regular replacement of certain POU system components.
- Improper monitoring and inefficient operation of POU units can result in non-compliance with drinking water standards. With a transition from centralized treatment to a POU system, control of the treatment process is dispersed from a central point to multiple points; thus to maintain the same level of control, more monitoring must take place.
- Application of POU devices for compliance by small systems with the arsenic rule can result in appreciable cost savings for the small systems, when compared to the cost of centralized treatment. Community water systems may be able to negotiate pricing arrangements with POU manufacturers that would provide appreciable discounts from retail prices. Sampling costs, however, can be much higher for multiple point testing of POU systems than single point testing of central treatment.
- Initial and continued education of the customers must be considered; new customers must be made aware of the system, and existing users must be reminded periodically of their responsibilities. In most cases, customers like and accept in-home treatment units and are satisfied with the systems and the water produced by POU units.

- Microbial constituents may be found in POU or POE treatment effluent. In particular, higher numbers of HPC bacteria may be present in POU effluent than in POU influent water. This may be reduced by periodic disinfection of POU/POE spigots or brief flushing of the spigots before use.
- Other problems observed in previous studies include sabotage, vandalism of the units due to a transient population, lack of utility personnel expertise to manage and coordinate sampling and maintenance, and use of other taps (by customers) for drinking water.

Regulatory Considerations

USEPA has approved the use of POU and POE treatment devices to achieve compliance with MCLs established in the National Primary Drinking Water Regulations, but certain restrictions have been placed on their use. These restrictions pertain to the ownership, control, and maintenance of the POU and POE devices, coverage by and ability to meet ANSI/NSF standards, equipment of POU and POE devices with alarm or shut-off mechanisms that are triggered when the device reaches a set capacity, and treatment of microbial and volatile contaminants and radon. Guidance is available from USEPA on selection of POU and POE technologies, installation, maintenance, and monitoring of the equipment, and information on costs associated with implementing a POU or POE treatment program and potential funding sources are provided.

Water systems considering implementing a POU or POE treatment program for SDWA compliance must consider state regulations and policies, which vary considerably from state to state. Some states (e.g., Pennsylvania) do not allow the use of POU devices for compliance. Other states allow the use of POU and POE devices for compliance with MCLs for only certain contaminants (e.g., arsenic, radionuclides, VOCs). A survey of state regulatory agencies indicated that several states have plans to develop policies and guidance for using POU/POE systems for SDWA compliance.

Performance

POU RO treatment failed to remove arsenic (III) at the Unity, Maine site and demonstrated that POU RO is not suitable at this site without pre-oxidation. POU Mn-AA was effective in removing As(III) in the Unity, Maine field screening evaluation.

In field screening evaluations, all POE units demonstrated excellent performance, treating over 50,000 gallons before breakthrough, except the POE Fe-AA unit at Stagecoach, Nev. Varying levels of iron in the raw water at this site may have caused plugging of the column and reduction of the media adsorption capacity. Periodic backwashing of the Fe-AA column would be required for successful POE application at this site.

Based on the results of the POU field screening evaluation at six sites, RO, GIM, AA, and Mn-AA were determined to be the most suitable technologies for in-home testing.

In-Home Testing

RO and Mn-AA POU treatment devices demonstrated overall excellent performance in the in-home testing program, removing arsenic to below detection limits of <0.002 mg/L even after 12 months of continued usage. Only two Mn-AA units demonstrated arsenic removals above detection limits; however, despite arsenic leakage of approximately 5-6 ppb, these units still

Table 9.1
Cost assessment summary

Market pricing	Monthly cost/household for POU RO		Monthly cost/household for POU adsorption	
	Scenario 1	Scenario 2	Scenario 1	Scenario 2
Low	$24	$21	$22	$20
Average	$31	$28	$26	$25
High	$40	$38	$29	$30

removed arsenic to levels below the 10 ppb MCL. From the statistical assessment of the in-home POU field testing results, it appears that a conservative POU monitoring frequency may be one sample per household per year. A reduced monitoring frequency of sampling only one-fourth of the total number of households per year, accompanied by adsorption media and membrane replacement frequency of once and twice a year, respectively, may also be acceptable. Optimal monitoring frequencies will vary between communities depending on raw water quality characteristics, seasonal variations in demand, and other site specific issues such as customer acceptance and participation, and state regulations.

Flow totalizer readings indicated that water usage between the participating homes varied considerably (from 338 to 1,200 gallons) over the course of the study. The majority of the homeowners, even those who minimally used the POU treated water, responded favorably with respect to using POU devices and the sampling program. Most homeowners experienced an improvement in the taste of the POU-treated water. Accessing the homes for monitoring and maintenance purposes was generally not a problem for utility personnel. Not surprisingly, there were some incidences where utility personnel experienced problems in coordinating convenient times for sampling and maintenance with the homeowners.

Costs

Based on the responses received for the survey of POU equipment vendors, the costs for RO and adsorption POU equipment, installation, replacement parts (e.g., filter cartridges, membranes), and service calls can vary considerably. Costs for RO POU units can range from $200 to $1,195, and adsorption equipment costs can range from $125 to $350. Considering the wide range of costs, water systems that plan to implement a POU or POE treatment program should obtain bids from several vendors, if possible, and references from each. When obtaining bids from vendors or manufacturers, water systems should investigate the possibility of obtaining a volume discount and a discounted service rate.

Table 9.1 summarizes the results of the cost assessment, with three market pricing conditions. Scenario 1 uses less frequent maintenance and more frequent monitoring while Scenario 2 uses more frequent replacement and less monitoring. The range of values in the table provides a general overview of expected POU costs. The above costs are similar, whether a utility purchases and monitors the POU units, or if the service is provided bu a vendor on a lease basis.

A comparison of central treatment costs for throw-away adsorption media and average POU treatment costs shows that POU RO and POU adsorption annualized costs were significantly lower (35-45%) than central treatment annualized costs for systems serving fewer than 50 connections. As the number of connections increased, the POU costs also increased gradually, and the breakpoint cost was observed at approximately 120 connections for POU RO and at approximately 200 connections for POU adsorption. Overall, POU treatment is cost-effective when compared to centralized treatment and is viable for small systems.

POE treatment costs are generally higher (39-48%) than centralized treatment costs for systems serving 20 to 300 connections. Therefore, POE treatment may not be considered a feasible option for small water systems, except in cases where savings in treatment by using POE versus central treatment are realized due to non-drinking and a significant amount of non-residential water uses (greater than two-thirds of total system demand).

Potential Barriers to Implementation and Solutions

Development and implementation of POU or POE treatment programs is complex and involves regulatory, utility, customer, liability, and other issues. Regulatory agencies are faced with addressing issues such as treatment of non-residential taps, 100% household participation, health risks associated with use of non-treated taps, required frequency of POU/POE monitoring and maintenance, disposal of wastes, and administrative burden and costs. Some of these issues may be addressed by requiring ordinances that make use of community water dependent on installation of a POU or POE unit, or by requiring agreements between water systems and customers to allow access to homes to maintain the POU or POE units. Water systems must address issues including household participation and retention in a program, treatment of non-residential taps, and gaining access to homes for installation, monitoring, and maintenance purposes. Customer related issues play an important role in the development of a successful POU or POE treatment program. These issues include customer acceptance, preference of a POE over POU device, and, in the case of a POU treatment program, use of untreated taps for consumption and cooking purposes. Liability issues such as product liability and liability associated with entering homes for installation, monitoring, and maintenance purposes may be addressed through adequate liability insurance and contracting services with licensed vendors. Customer education and communication are key to overcoming many of the barriers to implementation of a successful POU or POE treatment program. Effective outreach will help in developing a partnership between regulators, water systems, and customers.

RECOMMENDATIONS FOR FURTHER STUDY

An extended study at the in-home testing sites with additional sampling of the POU units would assist in estimating the exhaustion capacity of the RO membranes and adsorption media. This would provide valuable data that could be used to develop further recommendations regarding optimal POU monitoring and maintenance frequencies. Additional testing at the Unity, Maine site could provide further insight into the performance capabilities of POU units to treat arsenic (III). Several POU treatment technologies should be evaluated to determine the impact of treating arsenic (III) at this site with and without pre-oxidation.

Several new POU and POE treatment technologies are currently available commercially or are under development. The performance of these technologies should be evaluated at a larger number of field sites with a wide variety of water quality matrices.

In addition, various sizes of water systems which have successfully implemented a POU or POE treatment program should be evaluated in order to document the time that each spends on management, installation, monitoring, operations and maintenance, and reporting activities. This information could be used to provide guidance and develop accurate estimates of costs and staffing requirements for other water systems that are evaluating the feasibility of implementing their own POU or POE treatment program.

APPENDIX A
STATE REGULATORY AGENCY SURVEY

State Regulatory Agency Survey

	Contact Information	Date:	
		Name:	
		Title:	
		Street Address:	
		City, State:	
		Zip Code:	
		Tel:	
		Fax:	
		e-mail:	
Please enter responses within the green shaded/red bordered areas provided to the right of each question below.			
NUMBER	QUESTION	POSSIBLE RESPONSES	RESPONSE
1	What state agency has primacy authority for development of rules, policy and guidance concerning the use of POU/POE devices by public water systems for compliance?		
2	Do you have any of the following in place (or in development) on implementation, reporting, state oversight and compliance determination for POU/POE systems used for SDWA compliance? Place a Y in the green box next to the appropriate responses - more than one box may be marked.	POU rules	
		POU policy	
		POU guidance	
		POE rules	
		POE policy	
		POE guidance	
		None in place	
3	Does the state have in place (or in development) rules, policy or guidance that establish criteria regarding monitoring, including sample collection and location selection, for POU/POE systems used for SDWA compliance? Place a Y in the green box next to the appropriate responses - more than one box may be marked.	POU rules	
		POU policy	
		POU guidance	
		POE rules	
		POE policy	
		POE guidance	
		None in place	
4	If your state currently has experience in regulating POU/POE devices for compliance, identify limitations to current practices and potential solutions.	Limitations	
		Potential solutions	

5	What guidance and procedures are in place if a segment of the community does not want to install POU units in their homes (such as relief of regulatory burden for the utility for non-cooperation of certain homeowners, etc.)?		
6	Mark the boxes of the contaminants for which POU/POE systems are allowed (not including aesthetics) in your state. Place a Y in the green box next to the appropriate responses - more than one box may be marked.	Radionuclides	
		Arsenic	
		Chromium	
		Turbidity	
		Microbial	
		VOCs	
		Other (please list)	
7	What system sizes can utilize POU devices by regulation or policy? Place a Y in the green box next to the appropriate responses - more than one box may be marked.	None	
		<25	
		25 - 100	
		101 - 500	
		501 - 3300	
		>3300	
8	What are the size of systems that are currently using POU devices? Place a Y in the green box next to the appropriate responses - more than one box may be marked.	None	
		<25	
		25 - 100	
		101 - 500	
		501 - 3300	
		>3300	
9	If POU systems are currently used for compliance purposes in your state, please provide information on the system sizes, sources (GW or SW), contaminant, and treatment processes utilized:		
10	If POE systems are currently used for compliance purposes in your state, please provide information on the system sizes, sources (GW or SW), contaminant, and treatment processes utilized:		
11	If POU/POE systems are in use in your state, do you have any experiences or policies relating to water quality degradation in POU/POE systems due to the following: (place a Y in the green box next to the appropriate responses - more than one box may be marked)	membrane fouling	
		microbial degradation	
		loss of adsorptive capacity	
		other (please list)	

12	For communities with POU systems in use, please give an estimate of the % that discharge their wastewater to the sewer and the % that discharge their wastewater to septic systems.	% sewer		
		% septic system		
13	Are there concerns regarding the wastewater (concentrated brine) from POU reverse osmosis systems and its disposal?			
14	If applicable, what has been the attitude and perceptions of the consumers in the community where systems have been installed? Place a Y in the green box next to the appropriate responses - more than one box may be marked.	excellent		
		good		
		average		
		fair		
		poor		
		unknown		
15	Who is responsible for the operation and maintenance of POU/POE devices? Place a Y in the green box next to the appropriate responses - more than one box may be marked. List % of each if more than one.	Utility		
		Third party contractor		
		Equipment vendor		
16	What % of the POU/POE systems comply with SDWA MCLs? Place a Y in the green box next to the appropriate response - mark only one box.	>90%		
		>75-90%		
		>50-75%		
		>25-50%		
		£25		
17	What % of the POU/POE systems meet monitoring requirements? Place a Y in the green box next to the appropriate response - mark only one box.	>90%		
		>75-90%		
		>50-75%		
		>25-50%		
		£25		
18	Identify any planned or proposed activities your state has for SDWA compliance using POU/POE systems (i.e. arsenic, radionuclides):			
19	If you are aware of any local or county regulations that may exist within your state in the above subject areas, please identify the regulation, applicability, effective location(s) and provide contact information for the local regulatory agency.			

APPENDIX B
STATE REGULATORY DOCUMENTS

Requirements for Point of Entry (POE) Systems on Constructed Conveyance Water Systems

(Irrigation Districts)

California Department of Health Services
October 2000

TABLE OF CONTENTS

1.0 INTRODUCTION

Effective August 6, 1998, revisions to the federal Safe Drinking Water Act (Section 1401(4)(A)) expanded the definition of a public water system (PWS). The new definition states that an entity delivering water to the public through a pipe or "constructed conveyance" is considered a PWS if 15 or more connections are receiving and using the water for domestic purposes. Such systems may apply for an exemption from the PWS classification by either supplying the customers with an alternative water supply or by installing and maintaining point-of-entry (POE) treatment devices approved by the Department. The agency operating the constructed conveyance system is responsible for delivering the alternative water and/or overseeing the maintenance and operation of POE devices.

The purpose of this document is to provide guidance for irrigation districts and other entities that have recently been included in this expanded definition of a PWS that may want to utilize POE devices to provide customers with "equivalent water" and thereby qualify for an exemption. This document does not provide guidance for agencies intending to receive an exemption by delivering an alternative water supply, such as bottled water or hauled water. These requirements are outlined in a separate document.

The U.S. Environmental Protection Agency (EPA) has published a guidance document, dated July 31, 1998, on the definition of a public water system under the 1996 Safe Drinking Water Act amendments. The EPA guidance document states that the words "equivalent level of public health protection" are meant to distinguish the situation of providers covered by this section from the situation of public water systems which must comply with all relevant aspects of the applicable regulations, including sampling and testing requirements and sometimes details of treatment. For example, a point-of-entry treatment device for filtration and disinfection might not comply with all requirements of relevant drinking water rules for monitoring, extent of surveillance of the disinfection process, and so forth. But, it would meet the "equivalent level of public health protection" intent of this section if the quality of the water it produces is similar to that from central filtration and disinfection. Thus, this guidance sets performance criteria, attempting to assure that the quality of the water that affected residential users receive will be similar to that from central treatment.

The Department believes that POE devices used in accordance with the criteria outlined in this document are capable of providing "equivalent water" to that provided by an approved public water supply treatment facility. However, it is the opinion of the Department that the use of POE device treatment systems by individual customers of a constructed conveyance system may not provide a continuous safe, potable supply of water due to inappropriate operation and inadequate maintenance of these systems.

The Department will approve POE devices installed after July 1, 1999 using the same procedures it currently uses for the approval of alternative treatment technologies. Existing POE devices will be evaluated following the criteria outlined in Section 3.0 of this document. New POE devices which are identical to existing POE devices should not be installed at new locations without first complying with the criteria for "New POE Device Installations" as outlined in Section 4.0.

An irrigation district (ID) or other entity, including a third party consultant working with the constructed conveyance system, that is considering the use of POE devices by customers of their "constructed conveyance," should follow the guidance provided in this document. Prior to the Department's review of any new or existing POE devices, the constructed conveyance system should submit a plan to the Department indicating how the recommendations of these guidelines will be addressed.

2.0 CONSTRUCTED CONVEYANCE MONITORING AND OVERSIGHT REQUIREMENTS & RESPONSIBILITIES

The following general guidance applies to the constructed conveyance as a whole, and not specifically to individual POE devices.

2.1 System Responsibilities

Operators of constructed conveyance systems using POE devices are responsible for the following:

Ensuring that the devices installed have been reviewed by the Department.
Ensuring that the devices have been properly installed.
Ensuring that the devices are properly operating.
Conducting the recommended monitoring of the POE devices and constructed conveyance system.
Maintaining operation and maintenance records.
Maintaining records of the recommended monitoring results.
Providing the necessary annual report to the Department.
Providing education to customers on the operation of POE devices.
Ensuring that operators and maintenance workers who oversee the POE devices are properly trained.

2.2 Source Water Assessment

A source water assessment should be completed to assure that the constructed conveyance system, especially an agricultural ID, does not allow the blending of surface water with effluent from a wastewater treatment plant or any other significantly degraded water that would require additional treatment. This is especially important if the constructed conveyance is supplying water for domestic purposes.

2.3 Raw Water Monitoring Requirements

The monitoring guidance outlined in this section apply at the turnout to the constructed conveyance system and are not required monitoring of each individual POE device. The constructed conveyance operator responsible for the POE devices can use data from PWSs existing on the same constructed conveyance, provided the data is documented to be representative of the water being treated by the POE devices.

Turbidity - Turbidity should be monitored continuously and recorded by a data recorder at the turnout into the constructed conveyance. If water is added to the constructed conveyance at several points that results in significant changes in the constructed conveyance's raw water turbidity, monitoring at each of these points should be required.

Bacteriological water quality - Raw water total and fecal coliform density analysis should be performed weekly (MMO-MUG media should not be used).

Chemical and Radiological - Inorganic, organic, general mineral, general physical and radiochemical analyses should be conducted at the same frequency required of PWSs.

3.0 EXISTING INSTALLED POE DEVICE APPROVAL

The Department does not have the resources to review and approve all existing POE devices following the criteria outlined in Section 4 of this document. Therefore existing POE devices installed prior to July 1, 1999 can be considered as performing properly, provided the constructed conveyance system or a third party consultant acceptable to both the constructed conveyance system and the Department collects the data needed to document compliance with the performance criteria outlined below:

3.1 Membrane Filtration

All alternative technologies that use membranes and are currently approved by the Department have been assigned a 0.2 NTU performance standard. A 3-log *Giardia* removal credit will be granted to existing installed POE membrane devices if compliance with the 0.2 NTU turbidity performance standard is documented.

3.2 Bag and Cartridge Filtration

All alternative technologies that use bag and cartridge systems and are currently approved by the Department have been assigned a 0.2 NTU performance standard. A $1^1/_2$-log *Giardia* removal credit will be granted to existing installed POE bag and cartridge devices if compliance with the 0.2 NTU turbidity performance standard is documented.

3.3 Media Filtration

Existing POE media filtration devices, including diatomaceous earth and other granular media filters, should meet a 0.3 NTU performance standard. This is the performance standard set under the Interim Enhanced Surface Water Treatment Rule. A 2-log *Giardia* removal credit will be granted to existing installed POE media filtration systems if compliance with the 0.3 NTU turbidity performance standard is documented.

3.4 Inactivation

Existing inactivation devices should be capable of achieving the needed contact time (CT) inactivation criteria for the removal system with which they are installed. Ultra-violet (UV) inactivation credit will only be granted for virus. No *Giardia* inactivation credit will be granted for UV systems at the present time. If it is not possible to determine UV virus inactivation credit for an existing installed system, weekly coliform samples should be collected from the treated water after the UV treatment process for 3 months to determine if the unit is operating properly. Coliform-free water is needed to confirm virus inactivation credit.

Due to the limited data used to support the approval process of existing installed POE devices, the Department recommends that additional log inactivation be provided by the inactivation system if at all possible. This can be done by adding contact time, increasing the chlorine residual, or both for systems using chlorine. Systems that cannot meet the above outlined performance criteria should be replaced with a new approved device. The homeowner should be advised of the problems associated with the existing POE device and be advised to use bottled water until improvements have been provided.

4.0 NEW POE DEVICE INSTALLATIONS

The Department's Surface Water Treatment Rule (SWTR) Committee will review and approve the treatment capability of proposed POE devices following the same process and procedures used to review and approve an alternative treatment technology system (Policy # 94-002). To conserve the Department''s staff time required to complete a POE device review, the Department will not undertake a POE device review unless a constructed conveyance system has expressed a serious interest in the installation of the devices to be tested. Existing POE devices accepted under the criteria outlined in Section 3 should not be installed at new sites unless they have been evaluated for performance effectiveness using the criteria outlined in this section.

4.1 Treatment Performance Criteria

POE devices should be capable of achieving a total of 3-log *Giardia* cyst reduction and 4-log virus reduction. This level of treatment is consistent with the United States Environmental Protection Agency's (EPA) Surface Water Treatment Rule, promulgated in June 1989.

The Department may recommend that constructed conveyance systems delivering poor quality raw water provide a higher level of treatment. The Department will review raw water quality data and available watershed sanitary survey and source water assessment data before making that recommendation.

The Department expects POE devices to achieve the necessary *Giardia* log reduction through removal processes, and the virus reduction through a combination of inactivation and removal, or exclusively inactivation. Because of the difficulty in achieving >4-log removal of viruses, the Department envisions the use of a disinfection treatment process to inactivate viruses as a recommended treatment approach in conjunction with cyst and virus removal processes.

The companies developing these treatment devices should provide a failure analysis of the devices along with an estimate of the useful life of the various components included in the device, especially the parts that are critical to pathogen reduction. The analysis would include any component that requires periodic replacement to assure proper operation.

4.2 Potentially Acceptable Components of POE Devices

Alternative treatment technologies previously approved by the Department typically provide much larger flows than would be necessary for individual customers and are not generally considered to be POE devices, although the Department has approved one technology that could be used in this mode. However, there may be instances where an entity operating a constructed conveyance system may choose to install an approved alternative treatment technology for a cluster of customers or for an individual customer. If the cluster were to include 15 or more service connections or a population of 25 or more it would be considered a PWS, or if the cluster included 5 to 14 service connections, it would be considered a state small water system under local health department oversight.

The following treatment technologies are acceptable components of a POE device(s).

Passive Removal Technologies - The Department has approved a number of alternative technologies, including bags, cartridges, and membranes, that do not require chemical pretreatment of the source water. It may be acceptable to use these types of passive treatment technologies for POE purposes. Most of these approved technologies are capable of treating flows that are much larger than would be required of a POE device. However, it is likely that the technologies can be successfully scaled-down to a size more appropriate for a POE device. The cyst and virus removals provided by these technologies vary and an additional inactivation process is likely to be necessary in order to achieve the required total pathogen reduction.

Chemical Feed Technologies - Chemical feed equipment using solution feed pumps for pathogen inactivation are acceptable. The Department will also consider the use of erosion hypochlorination equipment. These types of equipment have a history of plugging due to scaling and providing widely varying levels of chlorine residuals and may not be appropriate for all sources. These systems should be equipped with a device capable of shutting down the POE device when the inactivation process has failed to perform properly.

UV Inactivation Technologies - UV light disinfection devices for virus inactivation will be acceptable provided they have been proven to be effective. They should be equipped with an internal light intensity probe with a visual indication of the relative intensity of the light source and be capable of shutting off the flow when the UV system is not functioning properly. Four-log UV CT values will need to be obtained from the EPA prior to the Department's final approval of any UV system claiming greater than 3-log virus inactivation. UV systems do not currently receive credit for cyst or oocyst inactivation. The recommended UV intensity necessary to achieve the required log virus inactivation will be established based on the data provided to the Department with the device approval documentation package.

4.3 Unacceptable Components of POE Treatment Devices

The following treatment technologies are not recommended as acceptable components of a POE device(s).

Chemical Based Removal Technologies - The Department does not recommend the use of POE devices that require chemical pretreatment to assure proper removal of pathogens. Changes in raw water quality would require continuous monitoring and adjustments that are not appropriate for a POE device. A high level of operational expertise would be needed to properly monitor and operate these devices. In addition, since a source water quality change is likely to affect all POE devices at the same time, the operator would need to simultaneously adjust the chemical pretreatment for all POE devices, a logistically impossible approach.

Gas Chlorination Technologies - The equipment needed to address the onsite safety concerns associated with this technology would make the use of gas chlorine too expensive to be considered practical as a POE component.

National Sanitation Foundation (NSF)-Approved Devices - The NSF assumes that their devices are to be used on approved public water supply systems. NSF makes no claims regarding virus removal capability. The NSF approved cyst reduction treatment devices are certified as providing cyst reduction on water supplies that are microbiologically safe but may contain cysts. NSF does not claim that they are capable of treating raw surface water and providing the appropriate level of pathogenic cyst reduction. Therefore, NSF-approved treatment devices should not be considered acceptable until additional testing is performed on raw surface water to demonstrate cyst and oocyst reduction.

5.0 MONITORING AND MAINTENANCE CRITERIA

The monitoring, maintenance and general POE oversight as outlined in this policy may be assigned to an independent third party acceptable to both the constructed conveyance utility and the Department. If a constructed conveyance utility should choose to provide central treatment of any kind it may request modification of its required monitoring. For example, if central disinfection were provided, the utility may present a system-wide treated water bacteriological monitoring program that is similar to the distribution system bacteriological monitoring required of PWSs. This may result in the reduction of some of the recommended monitoring of the individual POE devices.

5.1 Routine Monitoring

The Department will determine monitoring guidance for the POE devices on a case-by-case basis as not all POE device configurations will be the same. Some constructed conveyances may provide centralized disinfection, and POE devices on such a system may include only a cyst removal process at the customer's service. Monitoring and oversight of any central treatment should be identical to those required of PWSs with the same type of treatment. In general, the recommended monitoring will be as follows:

Turbidity Monitoring - As a minimum all POE devices should be monitored for treated water turbidity at least once every two weeks. The turbidity performance standard for each approved POE device will be established by the Department's SWTR Committee based on the data submitted by the POE device manufacturer. If the POE device contains a component such as a bag, cartridge or membrane that requires periodic cleaning and/or replacement at an interval that is shorter than 14 days, turbidity monitoring should be provided at an increased frequency. The individual turbidity performance standard will be set by the Department's SWTR Committee based on the data gathered in the manufacturer's approval demonstration study.

Chlorine Residual Monitoring - Each POE device that utilizes on-site chlorine addition for cyst and/or virus inactivation should be monitored for chlorine residual during each visit to the unit, or at least once every two weeks. The individual POE device chlorine residual performance standard should be based on the necessary CT values needed to achieve the required log inactivation.

Coliform Monitoring - The treated water from each POE device should be monitored for coliform bacteria on a monthly basis.

UV Monitoring - The relative intensity of each UV unit should be recorded by the system operators each time a visit to the POE site is conducted.

5.2 Follow-up Monitoring, Maintenance and Notification

If routine monitoring of the individual POE device detects a failure to comply with a performance standard, the homeowner should be immediately notified and corrective maintenance performed on the failing treatment process. The homeowner notification should include the precautions that must be taken by the homeowner to assure illness does not result. Boiling all drinking water or the use of bottled water for drinking are two precautions that homeowners can take that are recommended by the Department.

If the POE device cannot be made to perform properly within 24 hours, the homeowner should use bottled water until the problem has been corrected and follow-up monitoring demonstrates that the POE device is performing properly. After the corrective measures have been completed and the device is working properly, the constructed conveyance system should perform additional follow-up monitoring the next working day to assure the problem does not immediately reoccur.

5.3 Routine Maintenance Program

A routine operational maintenance program should be developed for all POE devices. The Department will review the proposed treatment device operations and maintenance program as part of the approval process. The operations and maintenance program should include the type and frequency of maintenance procedures that are needed to assure continued proper device performance, such as those identified by the POE device manufacturer, in addition to the performance monitoring requirements previously outlined. The maintenance program should also include procedures to adjust the performance of POE devices that are not meeting the performance criteria. The necessary operations and maintenance program will need to be developed prior to Department approval of a device. Constructed conveyance systems are expected to implement the recommended maintenance program using adequate staffing.

If a treatment device contains a component with limited useful functioning life that is critical to the proper performance of the device, a schedule for the replacement of this component should be developed. The replacement schedule should provide some safety factor applied to the known useful life that takes into account the fact that some parts are going to fail sooner than normally expected.

6.0 OPERATOR QUALIFICATIONS

Proper sample collection and analysis techniques are necessary to ensure that POE devices are operating properly. Therefore, it is critical that constructed conveyance staff or third party operator(s) who has the responsibility of overseeing POE operation, be properly trained in the operation and maintenance of the devices, collection of samples and analysis of appropriate water quality parameters (turbidity, chlorine residual, etc.). Treatment plant operators certified at Grade II or above can be considered properly qualified to conduct the required monitoring.

7.0 CUSTOMER EDUCATION AND OPERATIONAL AGREEMENT

The Department recommends that Constructed conveyance systems develop an educational program or brochure for the POE device users to educate them on the hazards of consuming improperly treated water. It should also include warning signs that might indicate improper POE device operation and the phone number to call if a problem is suspected.

At the time the POE device is installed, staff of the constructed conveyance system should meet with the individual customer to review the educational material with them. At the same time information should be provided outlining the ongoing monitoring and maintenance that will be performed on the POE device.

Personnel responsible for regular maintenance and monitoring of the POE device should have appropriate access to it. This may require a formalized written agreement between the homeowner and the constructed conveyance system. In addition, to ensure proper operation of the device, it is recommended that the customer not make any adjustments to the device without prior approval and guidance from the personnel assigned to oversee the POE device. The Department recommends that the devices not be accessible by the customer unless they have signed an agreement that stipulates the types of operational adjustments that they may make.

8.0 ANNUAL REPORTING REQUIREMENTS

The Department recommends that constructed conveyance systems keep appropriate records and provide an annual report to the Department that addresses the following issues.

- Customer identification, name, number in household, type of water service (well, bottled water or type of POE). Any change in status of a customer must be properly tracked and records kept of the changes that occur, including date of changes.
- Date of individual POE device monitoring and monitoring results (turbidity, disinfectant residual and coliform sample results).
- Complaints received.

- Investigations and actions taken to correct improper POE device performance.
- Training and certification of personal responsible for the POE device monitoring and maintenance.

9.0 LIST OF DEPARTMENT APPROVED POE DEVICES

The Department's approved POE devices will be listed as a subsection of the Department''s listing of approved alternative treatment technologies, which are on the Department's website. The website address is listed below:

(www.dhs.ca.gov/ps/ddwem/publications/pubindex.htm)

The above listed report is titled "California Surface Water Treatment Alternative Filtration Technology Demonstration Report" and can also be requested by mail at the address listed below.

Drinking Water Program, MS 92
P.O. Box 942732
Sacramento, CA 94234-7320

Attention: Joyce Silva

Guidelines for Approving Managed Point-of-Use Treatment in Idaho Public Water Systems

Idaho Department of Environmental Quality
October 2000

TABLE OF CONTENTS

INTRODUCTION

Point-of-Use (POU) devices treat water at a single tap. Point-Of-Entry (POE) devices treat water at the entrance to a home or other service connection. The 1996 Amendments to the Safe Drinking Water Act (SDWA) removed a long-standing prohibition on the use of these treatment methods for compliance with certain national drinking water standards. Idaho modified its regulations in 1999 to be consistent with the SDWA provisions.

All of the water within a facility served by POE treatment is suitable for drinking and cooking (although irrigation flows and other high-volume uses may be excluded from treatment using plumbing by-passes). However, because of cost and waste disposal issues, EPA does not consider POE treatment to be acceptable technology for small systems at the present time.

To clarify: The SDWA allows POE *and* POU, but EPA's Compliance Technology List (see references at the end of these guidelines) excludes POE in small system applications for the reasons listed in the preceding paragraph. The technology list does not carry the weight of regulation, so states are free to consider POE applications. In general, Idaho follows EPA's lead on these matters because DEQ lacks the staff to conduct evaluations of POE installations to weigh costs and benefits. In some situations, such as small transient systems, POE may make better operational and economic sense than POU. In these situations, the cost and waste management issues will be a deciding factor.

In some situations, the distinction between POE and central treatment becomes blurred. As a rule of thumb, central treatment is applied to all water produced by the system, whereas POE applications do not treat water used for irrigation or other non-potable purposes.

These guidelines deal primarily with POU applications.

POU treatment may be a safe and practical application for small systems under some circumstances, but there are two major concerns that must be addressed:

1. the treatment device is located on private property (usually in a home or business), which restricts the timing and convenience of access by the water system for the purpose of servicing the units.
2. not all water taps are treated to drinking water standards and the water system cannot prevent individuals from ingesting untreated water.

The 1996 SDWA attempts to address these concerns by requiring the public water system to be responsible for ownership and maintenance of the POU devices and by requiring that the devices be equipped with an alarm to notify customers of the need for maintenance or repairs.

These guidelines will provide technical background on POU treatment and offer guidance for DEQ engineers who are called upon to approve POU applications in Idaho public water systems. Water systems and their consultants should find information here to aid in preparing an engineering study. It should be noted that POU technologies capable of safely and reliably treating drinking water are available and well proven. For this reason, it is unlikely that the technical aspects of a POU application will be unusually challenging. **The primary burdens on the water system and their engineering consultant will be first, to make it clear to DEQ that conventional, centralized treatment is not feasible technically and/or financially, and second, to demonstrate with certainty that a management structure is in place to service and maintain the treatment equipment and monitor for compliance.** The normally passive role of the customer must be altered in order for POU treatment to succeed. The homeowner or business person must become an active and willing partner with the water system. This involves legal as well as educational arrangements. Managerial issues will constitute a primary focus of this document and will no doubt pose the foremost challenge to successful use of these treatment technologies.

SECTION 1. REGULATORY BACKGROUND AND GENERAL REQUIREMENTS

Regulatory Background

A public water system is responsible for delivering water that meets national drinking water standards to all of its customers. To obtain DEQ approval for POU treatment, it will be necessary for the water system to have written permission from all of its customers to allow water system personnel to enter private premises on a scheduled basis to install, maintain, or replace the POU units, and for the purpose of collecting compliance samples.

The SDWA states that ". . .point-of-use treatment units shall be owned, controlled, and maintained by the public water system or by a person under contract with the public water system to ensure proper operation and maintenance [O & M] and compliance with the maximum contaminant level or treatment technique." The use of POU treatment "to achieve compliance with a maximum contaminant level or treatment technique required for a microbial contaminant (or an indicator of a microbial contaminant)" is explicitly forbidden. Each treatment unit must be equipped with an alarm which alerts the customer to the fact that the unit is in need of attention and may not be producing safe water. Lastly, POU units must be independently certified according to any existing standards promulgated by the American National Standards Institute. Information is provided in Section 2 on alarms and on national standards that apply to POU devices.

General Requirements-Protocol for Submission and Approval

The following elements must be addressed in plan and specification submittals involving POU treatment:

1. **DEQ will only consider approving POU treatment upon receipt of an analysis which clearly demonstrates that central treatment is not feasible for the water system.** Central treatment is preferable to POU for safety and liability reasons. Unless waived by DEQ, this analysis must be in the form of an engineering report prepared by a professional engineer registered in the State of Idaho.
2. **Each water system customer must sign an access agreement which allows water system personnel to enter their property on a scheduled basis to install and maintain the treatment devices.** The agreement should include a statement that the homeowner assumes responsibility for instructing family members or other residents that they should not cook with or ingest water from untreated taps. The water system must

provide its customers with educational materials concerning the health risks associated with the contaminant(s) for which treatment is proposed. DEQ can provide basic health related information for the contaminants most likely to generate interest in POU treatment in Idaho. A termination of service clause/ordinance must be legally adopted by community systems in order to deal with uncooperative customers. In homeowner's associations, the access agreement and other informational materials associated with the community's water treatment strategy must be disclosed to all potential buyers so that cooperation by incoming residents who were not part of the original deliberative process may be assured.

3. **A pilot study may be required, particularly if multiple treatment units are planned, in order to determine the suitability of the POU device for treating the system's source water.** The scope and duration of the pilot study will be determined by such factors as the characteristics of the raw water, manufacturer's ratings of the treatment device, and good engineering practices. The pilot study will generate data on service intervals, aid in specifying and calibrating alarm systems, and reveal any site specific problems with component fouling or microbial colonization.
4. **The system must provide an operations and maintenance plan which demonstrates that the POU treatment units will be serviced in accordance with the manufacturer's instructions and that compliance sampling will take place, as required by DEQ.** Required service intervals will be determined initially by the pilot study or by following manufacturer's recommendations. The maintenance schedule can later be adjusted to reflect actual experience.
5. **The alarm mechanism that will be used must be specified in the engineering report.** See Section 2 for technical details.

Types of Water Systems

POU treatment may be more appropriate for some types of water systems than it is for others. The following discussion conveys DEQ's assessment of how POU applications should be approached in various system types.

1. New Systems: As a general principle, DEQ will resist POU applications in new systems. It is felt that with proper attention to well siting and construction it should be possible to avoid tapping contaminated waters that will require treatment. If uncontaminated waters are not available, POU will be considered on the same terms as existing systems.
2. Existing Systems: There are four categories of water systems with differing operating environments that can effect the suitability of various treatment options. In each case, the approval process will be in two parts. The first part is the system's justification, usually prepared by an Engineer. The second part is DEQ's staff analysis describing the agency's rationale for approving POU. Both parts must be carefully documented in the system file.
   A. Transient Systems: Included here are small bars and roadside stores or restaurants. DEQ will approve POU treatment where supported by practical and economic considerations (see general requirements, above). Nitrate is the only contaminant presently of regulatory interest to this group of systems, although systems in this

category may be interested in treating for other contaminants on a voluntary basis. In this environment, it is likely that the owner/operator of the system will service and maintain the treatment units. DEQ may consider plans and specifications prepared by a vendor. *Permanent* signs identifying non-treated water taps must be provided (for example, in restrooms). All water intended for public consumption, including ice and mixed beverages, must be treated. DEQ will consider a POE strategy in applications of this kind, provided that irrigation flows are excluded from treatment, waste disposal issues are effectively managed, and corrosive waters are treated as required.

B. Homeowner's Associations: POU treatment may be challenging in terms of access for repair or maintenance, particularly in the increasingly common circumstance where a subdivision is made up of recreational or part-time homes that are often unoccupied. These organizations may also lack police powers to aid in enforcing access and other forms of cooperation. If POU is allowed, DEQ will expect a clear demonstration, of universal homeowner support (see item 2 under General Requirements).

C. Non-Transient Non-Community Systems: POU may be a reasonable option for vegetable packing sheds, industrial plants, and other facilities with a large differential between potable and other water uses. Schools and day care centers may also qualify, although in each of these situations multiple taps may require treatment and costs of POU could quickly become comparable to central treatment. Rigorous attention must be given to the separation of potable and non-potable water systems and provision of an ongoing cross-connection control program.

D. Larger Community Systems (>~70 connections): POU can be an acceptable strategy where only a few homes require treatment, or when treatment of large volume of irrigation water is cost prohibitive. As previously stated, DEQ will require a feasibility study demonstrating the financial and practical advantage of POU over centralized treatment.

SECTION 2. TECHNICAL CONSIDERATIONS

This Section is based on sources that are reasonably current. However, technology continues to evolve, and periodic revisions of this information will no doubt be required.

Characterization of Source Water

The raw water to be treated must be characterized as appropriate for the technology being considered. Manufacturers will ordinarily specify raw water characteristics for which their equipment is suited, and this information may be used to determine the scope of raw water chemistry studies. As a minimum, inorganic constituents, including total dissolved solids (TDS) and pH, must be reported. If significant seasonal variations are suspected or known, a sampling scheme may need to be developed in order to capture the range of raw water parameters. Treatment choices should be based on worst case data. See also the checklists below.

Types of POU Treatment

There are currently two types of POU treatment approved by EPA for use in systems serving fewer than 10,000 customers. These are ion exchange, and reverse osmosis. The following table lists the treatment types and the contaminants for which they are approved. The bolded contaminants are those that are known to occur in Idaho and which are likely to be the best candidates for POU treatment.

POU Treatment	Contaminants
Ion Exchange (IX)	Lead, **Arsenic**, Barium, Beryllium, Cadmium, Chromium, Copper, Thallium
Reverse Osmosis (RO)	Antimony, Lead, **Arsenic**, Barium, Beryllium, Cadmium, Chromium, Copper, **Fluorosis**, Selenium, **Nitrate***

* EPA has *not* listed RO as acceptable technology for POU treatment of nitrate, but it is widely used for this purpose and will be considered in Idaho. Special precautions are required because nitrate at concentrations above the MCL can pose an acute risk for infants.

Ion exchange passes water through a bed of selective resins which exchanges contaminant ions for harmless ions. Anion exchange selects for negatively charged ions, such as nitrate and sulfate. Cation exchange selects for positively charged ions. The latter mechanism is the type used in household water softeners. When the resin has become saturated with contaminant ions, it is recharged by backwashing with a concentrated solution that replaces the absorbed contaminants with harmless ions. The brine solution used in backwashing is normally drained through an air gap to a sanitary sewer or subsurface sewage treatment system. However, acid solutions used in anionic systems may not be suitable for direct discharge to sewers.

Reverse osmosis uses a semi-permeable membrane to allow passage of water while blocking the passage of contaminants. Adequate system pressure is necessary in order to force the water to move against a concentration gradient (it is this reversal of the natural osmotic gradient that gives the process its name). From one-third to half or more of the water forced against the membrane is rejected to waste. This low recovery rate makes centralized or POE treatment with RO quite expensive. In addition, RO-treated water is corrosive and usually requires pH adjustment prior to distribution. POU treatment may be a more practical application of RO technology because only a small stream of water intended for drinking and cooking is ordinarily treated. A typical POU installation will have a 5 micron cartridge prefilter to remove particulate, the RO unit itself, which is plumbed to treat a bypass flow from a cold water tap, a granular activated carbon post-filter to improve the flavor of the finished water, and a storage tank of 1.5-3.0 gallons capacity to ensure that water is available on demand. This treatment train serves a dedicated tap, which will often be the only potable water source at a service connection. The reject water from the influent side of the RO unit must be plumbed to the sewer drain with an air gap to prevent backflow. All of this can usually be installed beneath the sink in a household kitchen. If post-treatment components are made of appropriate materials, the corrosive RO-treated water will not require pH adjustment prior to consumption.

Technical Checklists

As mentioned in the introduction, the burden of demonstrating feasibility rests with the water system and its consulting engineer, where one is employed. Preparation of a POU application by an engineer will be standard practice for all but the smallest of systems, and even in these instances DEQ reserves the option of requiring an engineering report. DEQ reviewers will need to check the submitted materials for technical accuracy and satisfy themselves that managed POU treatment will work for the system. The following checklists are provided to assist in both the preparation and review of a POU submittal. The checklists are not all-inclusive, but they do address some of the key factors to be considered in POU treatment.

Checklist 1—Reverse Osmosis Applications

- **Raw Water Chemistry:** Include analysis of all inorganic constituents that have an identifiable importance to the specific treatment device being considered. Manufacturers will ordinarily specify the raw water attributes that must be characterized. Report total dissolved solids (TDS) and pH. TDS > 1000 PPM can limit membrane effectiveness and longevity. Pre-softening of such waters may be required. pH may effect dielectric relationships at the membrane pores.
- **Membrane Type:** Cellulose acetate membranes (CAMs) and thin film membranes (TFPs) are the common types. Most RO applications use TFPs because of superior contaminant removal rates at wide temperature and pH ranges and at lower system pressure. Pressures below 20 pounds/sq. in. will result in unacceptable performance. 40-60 pounds of pressure is a suitable operating range for RO.
- **Capacity:** Typical household POU units will treat 8-15 gallons per day, depending on system pressure and water chemistry. Consumers will need to use water judiciously. Use of automatic icemakers may not be practical, or may require treatment units with larger capacity. Businesses will probably need multiple units or larger units (See Section 3 of these guidelines for discussion of management issues).
- **Installation:** Licensed plumber or qualified vendor personnel. DEQ will consider installation by system personnel if adequate training is provided by the equipment supplier.
- **Operating and Maintenance:** In a typical under-sink installation, initial filter and membrane replacement intervals will be developed from analysis of water chemistry and results of the pilot study. A written O&M plan is required. The water system may contract with a third party to perform this service. See Section 3 for management considerations.
- **Insurance:** Water systems have been held responsible for damages to residences incurred during installation or as a result of leaks. Adequate liability insurance should be included in the engineering study and cost analysis.
- **National Certifications:** ANSI/NSF Standard 58 applies to RO treatment systems. ANSI/NSF Standard 53 (health effects) also applies.
- **Warning Devices:** For RO applications, the best alarm device is an in-line TDS meter. When breakthrough at the membrane occurs, TDS in the treated water will increase sharply. This can be used to actuate a warning light. The alarm mechanism must be checked for functionality and calibrated at each scheduled maintenance. For nitrate

applications, the TDS meter should actuate a shut-off valve to prevent ingestion of untreated water by infants.

- **Waste Streams:** Impacts of waste streams from POU units will generally be minimal. However, in larger communities the effect of these streams may need to be considered in terms of their impact on the operation of central sewage facilities. The engineering report should estimate the volume of the total waste stream and its contaminant load. This can be based on pilot study data regarding reject and treated water volumes and percent removal of source water contaminants. Businesses such as restaurants or others with large demands for potable water may require special consideration of waste impacts, especially if subsurface sewage disposal is practiced.
- **Disinfection:** Water delivered to the point of use device should ordinarily be thoroughly disinfected to prevent pathogens from entering the POU treatment train. Water systems with a record of high bacteriological quality may wish to pilot the POU equipment without disinfection and conduct evaluations of bacterial colonization in the treatment train. Disinfection is not mandatory, but may be required if bacterial problems arise after installation. The engineering study must consider this contingency. In systems that disinfect with chlorine, the residual should not be excessive or it may attack the membrane and result in impaired treatment efficiency and reduced lifespan. A carbon pre-filter may be needed. See Section 3 for a discussion of risks associated with bacterial quality in POU-treated water.

Checklist 2—Ion Exchange Applications

- **Raw Water Chemistry:** The raw water should be characterized for inorganic constituents and other attributes pertinent to the treatment equipment being considered. As with RO systems, the manufacturer will ordinarily specify the scope of the raw water chemical analysis that is need to determine the suitability of various treatment options. Competition among ions for exchange sites on the resin can seriously impact the efficiency with which the contaminant(s) of concern are removed. Manufacturers will have ion selectivity sequences for the resins in their products. Some waters may not lend themselves to IX treatment without prior softening. Magnesium, copper, and iron can foul IX equipment.
- **Resin Type:** Cation exchange is effective in removing barium, cadmium, copper, zinc, manganese, chromium III, iron II, lead, mercury, and radium. The sodium chloride (salt) solution used to backwash these devices is not harmful to sewage facilities. Anion exchange is effective at removing nitrate, bicarbonate, arsenic, chromium, selenium, sulfate, and chloride. Anion exchange units are backwashed with an acid solution. Handling and storage of the acid and disposal of the backwash solution can be problematic.
- **Capacity:** IX units are capable of treating larger volumes of water than RO devices. However, the size of the resin tank and other components is the limiting factor. In general, IX is viewed as being more suited to point of entry treatment, but cost and waste disposal issues have prevented EPA from recommending POE applications for small systems at the present time.
- **Installation:** A licensed plumber, vendor personnel, or adequately trained system personnel can install this equipment. The system or its consulting engineer must

consider local or state plumbing codes. Waste backwash solutions must be discharged to the sewer through an air gap to prevent backflow.

- **Operation and Maintenance:** Units must be checked on a regular schedule to insure that mineral fouling or bacterial colonization is not occurring. Salt or acid supplies must be regularly replenished. The timing of maintenance activities will be determined during pilot studies.
- **Insurance:** Liability and damage insurance must be maintained by the water system to cover losses to homeowners due to leaks or installation and maintenance activities.
- **National Certifications:** ANSI/NSF Standard 53 (health effects). ANSI/NSF Standard 44 applies to ion-exchange systems.
- **Warning Devices:** A flow meter which shuts off the water supply after a pre-determined amount of water has been treated is an effective alarm mechanism for ion exchange systems. The amount of water that can be treated before the unit must be serviced is determined during the pilot study. A safety factor is built into the flow meter shut-off setting. Alarm devices must be checked for proper function and calibrated or repaired as necessary at each scheduled service.
- **Waste Streams:** Not usually an issue with POU applications. Anion exchange devices, if proposed, will require an analysis of waste volumes and impacts.
- **Disinfection:** Water delivered to the treatment device will ordinarily require disinfection to prevent bacterial colonization or fouling of the treatment train.

SECTION 3. MANAGEMENT CONSIDERATIONS

The traditional model for water treatment in public systems involves a central treatment facility that treats all water sent into the distribution system for delivery to customers. The treatment plant is under the supervision and control of skilled operators who optimize treatment processes and observe rigid safety standards.

POU treatment departs radically from the traditional model. It distributes the treatment plant throughout the service area of the water system. It treats only a small portion of the customers' water, leaving the remainder untreated and potentially available to residents who may be unaware of the treatment strategy or simply forgetful. The unique risks and drawbacks associated with POU treatment have been emphasized in the introduction to this document.

In this Section, information will be provided to aid in designing a *managed* POU treatment strategy. The water system and its customers must form a partnership to ensure that people in the service area do not ingest untreated water. This involves a process of educating the community and then reinforcing that learning on a perpetual basis. It also involves regular and careful maintenance of the treatment devices to ensure that they do not fail.

Health Risks

In the technical section of these guidelines, it was suggested that arsenic, fluorosis, and nitrate are the drinking water contaminants that are known to occur in Idaho and are likely candidates for point of use treatment. Nitrate is a special case and is discussed below. When arsenic and fluorosis exist in drinking water in quantities exceeding the maximum contaminant level (MCL), they pose a *chronic risk* to those who drink the water. The MCL for fluorosis is 4.0 mg/l and for arsenic it is currently .05 mg/l. EPA has proposed reducing the arsenic MCL to .005 mg/l. Chronic contaminants at or greater than MCL concentrations can cause serious health consequences when ingested over a lifetime. Arsenic may cause skin damage, circulatory problems, and increased risk of cancer. Fluorosis may cause skeletal flourosis, a condition characterized by tenderness and pain in the bones. Children may experience mottling of their teeth.

Consuming a small amount of this water on an accidental or occasional basis will not be likely to cause harm. Under these conditions, a POU application may be a good solution to what otherwise could be an expensive treatment problem for a small water system. However, as concentrations of these contaminants increase significantly above the MCL, central treatment or provision of an alternative source of water become better solutions.

The nitrate MCL is set at 10 mg/l, a level that is protective of infants. Nitrate is converted to nitrite in the body and can inhibit the oxygen carrying capacity of the blood, resulting in a potentially fatal disorder known as "blue baby syndrome," or methemoglobinemia. Because nitrate is a potentially acute contaminant to infants, the risk associated with accidental ingestion of untreated water is much higher than it is for arsenic and fluoride. An active and continuous public education process will be required if POU is used to treat for nitrate. Customers must *never* make infant formula with untreated water.

Affordability

As a rule of thumb, costs for water in a community should fall between 1.5 and 2% of median household income. Obviously, a community may be willing (or may be required) to pay more than this if it has poor quality water that requires expensive treatment. Although central treatment is generally preferred because of safety and liability concerns, the capital and operating costs of this type of treatment may be beyond the reach of a small system. It may be difficult for a small system to attract and retain a qualified operator for a central treatment facility. Waste disposal may be expensive, particularly in isolated communities. Large quantities of water destined only for irrigation use must be treated to potable standards. These and other, similar challenges connected with central treatment will probably lead to increased interest in POU applications in small systems.

County level data on numbers of persons per household, per person income, and other demographic and economic factors that may be useful in a feasibility analysis are available from the Idaho Department of Commerce website: www.idoc.state.id.us/idcomm/cntypro.html.

Operation and Maintenance

As outlined in the regulatory background on page 5, the water system retains ownership of the POU treatment devices and is responsible for their routine maintenance as well as for the quality of water delivered by each unit. The following points may be useful when determining how to design and manage an O & M strategy:

1. Vendors of POU equipment may be interested in contracting for the routine service and repair of their products. This would be a favorable arrangement for small water systems that may lack qualified personnel.
2. Most service will need to be performed during evenings and weekends, since many people are not at home during normal weekday business hours. This could become a demoralizing factor for those charged with the maintenance task. It will no doubt add substantially to the cost of any service contract that may be negotiated, and it may make it difficult for a system to attract and maintain a qualified operator.
3. The customer must become a partner with the water system in detecting leaks, noting the status of the alarm device, and promptly communicating any other problems that may be encountered. This requires a perennial outreach activity that can be costly and demanding. Costs associated with public outreach are an important component of the feasibility analysis.

Microbial Concerns

If influent water is disinfected, problems with microbial build-up in the POU treatment train are likely to be minimal. However, due to the ubiquitous presence of bacteria in the environment, it will usually be difficult or impossible to avoid bacterial colonization of the treatment train over time. A granular activated carbon filter is probably the system component most susceptible to bacterial growth, since carbon serves as an energy source for many of these organisms. A GAC filter may be used prior to a RO unit to remove chlorine (as protection for the membrane) and again after treatment to polish the finished water. Several scientific studies have demonstrated that the types of bacteria that grow in GAC filters are not harmful to healthy persons, *provided that the water arriving at the treatment device does not contain pathogenic organisms.* If GAC filters are changed on a frequent basis, bacterial populations will not have time to build up to significant levels. Periods when water temperatures are warm would pose the greatest risk of microbial growth. Again, the key factor seems to be that the influent water must be of high microbiological quality. Post-treatment UV disinfection could be provided to reduce the risk of microbial contamination, but this has not usually been necessary with POU applications.

Compliance Monitoring

Systems will continue to be required to monitor for coliform bacteria at locations representative of water quality in the distribution system (not at the tap serviced by the POU device). In addition, these systems may wish to periodically sample for heterotrophic bacteria in POU-treated water to determine if filters are being changed frequently enough. DEQ will develop a chemical contaminant monitoring protocol that is sufficient to provide a statistical test on the adequacy of maintenance intervals. These samples should be taken at the time of scheduled maintenance, but prior to servicing the treatment unit. Regional and State Office staff will need to consult on the chemical monitoring scheme for each individual water system until enough experience has accumulated to make it possible to take a standardized approach. In theory, there should be no samples that exceed the MCL, because the POU devices must be equipped with an alarm device to signal treatment failure. See the technical checklists in Section 2.

Businesses and Other High Usage Connections

Communities that have the legal authority to pass ordinances have sometimes required businesses and other service connections that require larger quantities of potable water to purchase their own treatment devices. The water system specifies the equipment and retains ownership and responsibility for maintenance. Waste management can become a significant issue in these circumstances. If a community is of sufficient size to have a fair number of high usage connections, it is likely that the financial feasibility of central treatment will improve to the point that POU strategies will be rejected.

Advantages of POU Treatment for Small Systems

These guidelines have emphasized the limitations and disadvantages of POU treatment because these factors are crucial to the development of a safe and effective treatment strategy. Once these matters have been effectively considered, it may be encouraging to weigh some of the benefits associated with POU treatment:

1. It is not necessary to treat large volumes of irrigation water, which make up about 95% of the water delivered to customers in a typical water system.
2. Waste liquids or solids are produced in small quantities and usually may be disposed of in sewage treatment systems. Central treatment, by contrast, often produces a significant volume of waste sludges or brines which must be disposed of at considerable expense.
3. Involvement of the consumer in water treatment issues will, over time, foster a customer base made up of educated and involved people. This has the potential to make managing system operations and finances a much simpler task.
4. POU treatment is likely to encourage conservative use of potable water, in that the quantities produced by most home scale devices are quite limited.

Reporting and Recordkeeping

DEQ will expect water systems using POU treatment to maintain records of maintenance activities, public outreach, and costs of service. An annual summary of costs and performance data for the POU strategy should be available for inspection on request. Systems may wish to contract with their design engineer to prepare this annual summary.

Concluding Remarks

POU can be an effective treatment strategy for small systems. Reverse osmosis treatment for nitrate, arsenic, or fluorosis is likely to be the most common POU choice in Idaho at the present time. The management challenges associated with POU treatment are significant and must be considered by the water system on the same level of importance as technical issues. DEQ welcomes comment on these guidelines from consultants, water systems, and other users. These guidelines will be revised as regulatory and technological changes occur.

REFERENCES

Lykins, B.W. Jr., Clark, R.M., and Goodrich, J. A. *Point-of-use/Point-of-entry for Drinking Water Treatment*. Lewis Publishers, Ann Arbor, MI. 1992.

USEPA, 1998 *Cost Evaluation of Small System Compliance Options: Point of Use and Point-of-Entry Treatment Units*. Prepared by The Cadmus Group, Waltham, MA.

USEPA, 1998. *Small System Compliance Technology List for the Non-Microbial Contaminants Regulated before 1996*. EPA 815-R-98-002.

Commonwealth of Pennsylvania
Department of Environmental Protection
Bureau of Water Supply and Wastewater Management
Division of Drinking Water Management

CHAPTER 109. SAFE DRINKING WATER POE/POU REGULATIONS

Chapter 109 can be accessed at the following link: <http://www.pacode.com/secure/data/025/chapter109/chap109toc.html>

The following are the portions pertaining to POE and POU treatment devices.

§ 109.1. Definitions. *****

Point-of-entry (POE) device-A treatment device used as an alternative to central treatment that is installed on a public water line or service connection to a house, building or other facility for the purpose of reducing contaminants in the water distributed throughout the house, building or facility.

* * * * *

§ 109.301. General monitoring requirements. *****

(9) ***Monitoring requirements for POE devices***. A public water supplier using a POE device shall, in addition to the monitoring requirements specified in paragraphs (1)-(8), conduct monitoring on the devices installed. As a minimum, the monitoring shall include the MCLs for which the POE device is intended to treat and monthly microbiological monitoring. The Department may allow the water supplier to reduce the frequency of microbiological monitoring based upon historical performance. Except for microbiological contaminants, monitoring shall be performed quarterly on 25% of the installed POE devices with the locations rotated so that each device is monitored at least once annually, unless increased monitoring is required by the Department under § 109.302.

* * * * *

§ 109.303. Sampling requirements. *****

(g) A compliance sample required under § 109.301(9) shall be taken at a free flowing tap in the house, building or facility where the POE device is located or at a monitoring point approved by the Department on the effluent side of the POE device.

* * * * *

§ 109.407. General public notification requirements. *****

(c) ***Public notice recipients.*** *****

(3) If a public water system has a violation involving a point-of-entry (POE) device, the Department may allow the water supplier to limit distribution of the public notice to only persons served by that POE device. Permission for limiting distribution of the notice shall be granted in writing by the Department.

* * * * *

§ 109.503. Public water system construction permits. *****

(a) ***Permit application requirements.*** *****

(2) ***Special requirements for public water suppliers proposing to use POE devices.*** Permit applications which propose the use of POE devices shall, in addition to the information required in paragraph (1), include the following:

(i) Documentation that each POE device to be used meets the certification requirements of § 109.612 (relating to POE devices).

(ii) Manufacturer's design and engineering information, including blueprints or similar drawings, which provide detailed information about the construction and operation of the treatment device and its components.

(iii) A detailed monitoring plan, subject to the Department's approval, which includes a list of the contaminants to be monitored and the frequency of monitoring.

(iv) An operation and maintenance plan, as outlined in § 109.702 (relating to operation and maintenance plan), which includes a schedule of routine maintenance to be performed and the parameters to be monitored to determine the performance and condition of the devices.

(v) A drawing of the water supply distribution system showing each house, building or facility where POE devices are to be installed.

(vi) Proof of the right-of-access for every house, building or facility to be served by a POE device.

* * * * *

§ 109.602. Acceptable design.

(e) Point-of-use devices which are treatment devices applied to a single tap are not an acceptable treatment method for complying with an MCL or treatment technique requirement.

* * * * *

§ 109.612. POE devices.

(a) POE devices may be approved by the Department for use only by a public water supplier serving 100 or fewer individuals for the treatment of sources permitted prior to May 16, 1992.

(b) POE devices used by a public water supplier shall be tested and certified by the NSF or other certification organization acceptable to the Department against ANSI/NSF standards established for drinking water treatment devices. To be acceptable to the Department a certification organization other than NSF shall have a program at least as stringent as the NSF program and meet the requirements under § 109.606(d) (relating to chemicals, materials and equipment) as applicable to ANSI/NSF standards for drinking water treatment devices.

(c) A public water supplier using POE devices as a means of treatment shall install a POE device on the service line to customers, except for customers who are provided with water that meets the requirements of Subchapter B (relating to MCLs, MRDLs or treatment technique requirements) without the use of a POE device.

(d) The design, installation and operation of a POE device shall be of a type that the microbiological safety of the water is maintained.

* * * * *

§ 109.712. Control of POE devices.

(a) The public water supplier shall be responsible for control of POE devices installed under a permit according to § 109.503(a)(2) (relating to public water system construction permits). This includes the installation, operation and routine maintenance of each device.

(b) A public water supplier which installs a POE device shall obtain and maintain a right-of-access to the house, building or other facility where the POE device is installed in the form of a covenant running with the land.

APPENDIX C
ADSORPTION MEDIA AND RO MEMBRANE SPECIFICATIONS

ADSORPTION POU UNIT SPECIFICATIONS

Adsorption media	Cartridge	EBCT (min)	Flow (gpm)
AA or Mn-AA	4.5" × 20"	3.4	0.4
AA or Mn-AA	4.5" × 10"	1.7	0.4
GFH	4.5" × 20"	3.4	0.4
GFH	4.5" × 10"	1.7	0.4

Media and description	Particle size	Special preparation	Packed density (lbs/ft^3)	Adsorptive capacity (μgAs/g media)
Aquabind EP (alumina-oxide composite)	28 × 48	rinse fines	38-41 (dry)	500-1,000
Aquabind MP (manganese coated alumina-oxide composite	28 × 48	rinse fines	41-47 (dry)	500-2,000
Alcan FS-50 (iron-coated activated alumina)	28 × 48	rinse fines	45-50 (dry)	1,000-4,000
Granular ferric hydroxide (precipitated ferric salt processed into wet granules)	0.4-2.0 mm	rinse fines	82 (wet)	800-8,500

RO Thin-Film Membrane Specifications (per manufacturers' data)

Membrane type	Salt rejection (minimum)	Membrane area (sq. ft)	Flux @ 40 gpd (gfd)	Min. line pressure required (psi)	Max. operating pressure (psi)
Dow Filmtec Low Pressure TW30	96% (NaCl)	3-5	8-13	40	125
Desal AG (RO)	99% (NaCl)	3-5	8-13	40	125
Desal DK (NF)	98% ($MgSO_4$)	3-5	8-13	40	125

Unit Specifications for Sun City West, Ariz. Site

Unit	Flow (gpm)	Design criteria	Dimensions
POU - RO	0.4	Membrane type: Dow Filmtec	13.13" × 16.37" × 5.18"
POU - Aquabind MP	0.4	EBCT: 1.7	4.5" × 10"
POU - Aquabind EP	0.4	EBCT: 1.7	4.5" × 10"
POE - GFH (GEH Wasserchemie)	5.0	EBCT: 2.5 Media depth: 2.2 ft	12" dia. fiberglass tank
POE - Fe-AA (Alcan FS-50)	3.0	EBCT: 5.0 Media depth: 2.6 ft	12" dia. fiberglass tank

Unit Specifications for Carson City, Nev. Site

Unit	Flow (gpm)	Design criteria	Dimensions
POU - RO	0.4	Membrane type: Desal DK	13.13" × 16.37" × 5.18"
POU - GFH (GEH Wasserchemie)	0.4	EBCT: 1.7	4.5" × 10"

Unit Specifications for Unity, Maine Site

Unit	Flow (gpm)	Design criteria	Dimensions
POU - RO	0.4	Membrane type: Desal AG	13.13" × 16.37" × 5.18"
POU - Aquabind MP	0.4	EBCT: 1.7	4.5" × 10"

Unit Specifications for Metro Tucson Water, Ariz. Site

Unit	Flow (gpm)	Design criteria	Dimensions
POU - RO	0.4	Membrane type: Desal DK	13.13" × 16.37" × 5.18"
POU - Aquabind EP	0.4	EBCT: 3.4	4.5" × 20"
POE - GFH (GEH Wasserchemie)	5.0	EBCT: 2.5 Media depth: 2.2 ft	12" dia. fiberglass tank
POE - Fe-AA (Alcan FS-50)	3.0	EBCT: 5.0 Media depth: 2.6 ft	12" dia. fiberglass tank

Unit Specifications for Stagecoach, Nev. Site

Unit	Flow (gpm)	Design criteria	Dimensions
POE - GFH (GEH Wasserchemie)	5.0	EBCT: 2.5 Media depth: 2.2 ft	12" dia. fiberglass tank
POE - Fe-AA (Alcan FS-50)	3.0	EBCT: 5.0 Media depth: 2.6 ft	12" dia. fiberglass tank

Unit Specifications for Houston, Texas Site

Unit	Flow (gpm)	Design criteria	Dimensions
POE - GFH (GEH Wasserchemie)	2.0	EBCT: 2.5 Media depth: 1.9 ft	8" dia. fiberglass tank
POE - Fe-AA (Alcan FS-50)	2.0	EBCT: 2.5 Media depth: 1.9 ft	8" dia. fiberglass tank

APPENDIX D
POU/POE FIELD SCREENING PROTOCOL

Table D.1
POU/POE field screening site locations and arsenic concentrations

Site name	Raw water quality	POU/POE to be tested
Sun City West, Well 28B (Phoenix Metro Area), Ariz.	As^{+3} = ND As^{+5} = 31 ppb Silica = 28 mg/L	POU: RO, Mn-AA, AA POE: 3 to 5 gpm GFH, Fe-AA
Sandy Stream Water System (Unity, Maine)	As^{+3} = 99 ppb As^{+5} = 5 ppb pH = 7.69	POU: RO, Mn-AA
Well 3, Metro Water, Tucson, Ariz.	As^{+3} = ND As^{+5} = 12-14 ppb pH = 6.9 Silica = 42 mg/L	POU: RO, AA POE: 3 to 5 gpm GFH, Fe-AA
Stagecoach, Nev.	Total As = 29 ppb pH = 7.8 Iron = 1.29 mg/L Silica = 66 mg/L	POE: 3 to 5 gpm GFH, Fe-AA
Carson City, Nev.	As^{+5} = 15 ppb pH = 8.3 Silica = 17 mg/L	POU: Mn-AA, GFH

1.0 INTRODUCTION

This document summarizes the testing protocol to screen several POU/POE treatment units prior to installation in consumer's homes. Field screening was conducted at six different locations with varying water qualities to determine the impacts of water quality on the various POE/POU technologies. The six test locations are listed in Table D.1. The treatment processes that were tested include four types of adsorptive processes (Activated Alumina [AA], manganese coated AA [Mn-AA], iron coated AA [Fe-AA], and GFH), and reverse osmosis (RO) membranes with thin film (TF) membrane elements.

The field screening program consisted of two phases. Phase A spanned two weeks and included several on/off cycles followed by an eight hour rest period. Results from this phase were compared to results obtained from manufacturers on NSF challenge waters. For the POU units, the first two weeks of the schedule were designed according to NSF protocol, which stipulates operation of the units on 40-minute on to 40-minute off cycle, 16 hours per day, followed by an 8 hour rest period under pressure. Phase B, lasting 10 weeks, was designed to mimic actual household operation of the POU units and to aid in the subsequent selection of POU devices to be installed in houses in the following phase of the project.

This screening program was designed to address the following issues:

- Long term arsenic removal - arsenic was monitored for a 3-month period to determine the reliability of the POU units and aid in the design of the in-home testing phase of the study.
- Fouling issues - scaling of membranes due to inorganics, TDS and microbial fouling were monitored.
- Logistics and O&M issues - the labor effort required for sampling and maintenance of the units were evaluated.
- Microbial degradation - weekly sampling of total coliforms, Heterotrophic Plate Counts (HPCs), and ambient temperatures were recorded to assess microbial growth in the units.
- Arsenite removal - the potential of POU/POE units to remove As^{+3} was evaluated at the Unity, Maine site.
- Preoxidation - the effects of preoxidation on arsenic speciation and removal were examined at the Unity, Maine site.
- Intermittent operation and vacation periods - the units were operated to mimic daily use in houses and long vacation periods were simulated to observe the effects on the performance of the units.
- Post treatment - the potential for post treatment with GAC (used for mitigating taste and odor issues) to give rise to microbial colonization was examined.
- Flow indicator devices - the efficacy of the indicator devices on each of the POU RO units was evaluated. The RO units included a filter life indicator that, after 500 gallons have passed through the postfilter, stops the flow of water to the faucet. The cartridge must then be replaced to restore the water flow.

2.0 SAMPLING OBJECTIVES FOR POU/POE TECHNOLOGIES

The 3-month field screening evaluation provided an independent verification of POU/POE treatment unit performance for the processes tested. The POU adsorption (GFH, Mn-AA, Fe-AA, and AA) processes were tested using an EBCT of 1.7 minutes and 3.4 minutes and a POU flow rate of 2 liters per minute. POU adsorption cartridges with two different sizes (4.5" × 10" and 4.5" × 20") were used to achieve the appropriate EBCT. The 12" diameter adsorption column POE systems were tested using a 5 minute EBCT and a flow rate of 3 gpm for Fe-AA media and 2.5 minute EBCT and 5 gpm for GFH. Standard landscape sprinkler timers (120 volt) which can be programmed for up to nine start times per day were used to achieve the desired on and off periods for the units. Although the screening tests were conducted at higher daily average to instantaneous flow ratios than typical home use, the effects of intermittent operation were investigated and the units were operated until breakthrough was reached.

Equipment Specifications

The POU adsorption units were available in cartridges sizes of 2.5" × 10", 4.5" × 10" and 4.5" × 20". The 4.5" × 10" and 4.5" × 20" cartridges were used in the field screening phase of this project. Appendix C presents the various specifications for the adsorption media and cartridges which were screened at the field sites. The 2.5" × 10" cartridges were tested due to the low EBCT value. The POU cartridges also incorporated granular carbon blocks. These cartridges can be connected in to a bypass line from the cold water line. The unit has automatic shutoff after a certain amount of water has been treated.

Long-term rejection of arsenic in relationship to RO membrane fouling has not be proven in POU systems. The field verification will determine the impacts of microbial fouling on the membrane, and fouling due to accumulation of precipitates such as barium sulfate, calcium sulfate, and calcium carbonate. Since pH reduction or addition of antiscalants cannot be accomplished in POU systems membrane scaling and fouling will have to be quantified with respect to arsenic rejection.

The Kinetico RO POU unit consists of a prefilter cartridge, an RO membrane, a holding tank, a faucet, and a postfilter cartridge. Feed water is input to the prefilter. After prefiltering, water flows to the RO membrane. As treated water is produced, it is stored in the holding tank under pressure using a bladder concept. The more water that is stored, the higher the tank pressure is. From the holding tank, water flows to the postfilter and then is delivered for use to a faucet. The prefilter cartridge is a carbon/sediment cartridge for removal of sediment and chlorine prior to introduction to the membrane element. Three types of membranes are to be used: a thin film (TF) Dow Filmtec TW30 low-pressure RO element, Osmonics Desal AG RO element, and Osmonics Desal DK NF element. Membrane specifications are provided in Appendix C. The postfilter cartridge was a GAC cartridge used for improving tastes and odors. The RO units operated off of pressure supplied from the water source and did not require an external electrical connection.

3.0 TEST LOCATIONS AND SYSTEM CONFIGURATION

Sun City West, Well 28B, (Phoenix Metropolitan Area)

One POU RO unit, one POU Mn-AA unit, and one POU unit with AA media were installed at the Phoenix area well site and run on a timer based on the protocol described in the next section. Two 3 to 5-gpm 12-inch POE GFH and Fe-AA media columns were also tested at this site (both in intermittent and continuous mode operation). The schematic for this site is shown in Figure D.1.

Sandy Stream Water System (Unity, Maine)

At this site the arsenic is mostly in the +3 state. This gives us an opportunity to test the POU technologies for the As^{+3} species also. This represents a site that has a low level of competing reductants and is the site where Filox (solid preoxidant media) was first applied for As^{+3} oxidation. A treatment plant using Filox oxidation followed by anion exchange has been operating for a period of nearly 11 years. The current Filox bed is approximately three years old and the IX bed is the original resin system with an Ion Exchange unit is operational at this site. The POU units tested at this site will be installed prior to the Filox filter to assess As^{+3} removal by each of the technologies installed at the site. One POU RO unit and one POU Mn-AA unit will be set up at this well site and run on a timer based on the protocol described in the next section. The schematic for this site is shown in Figure D.2.

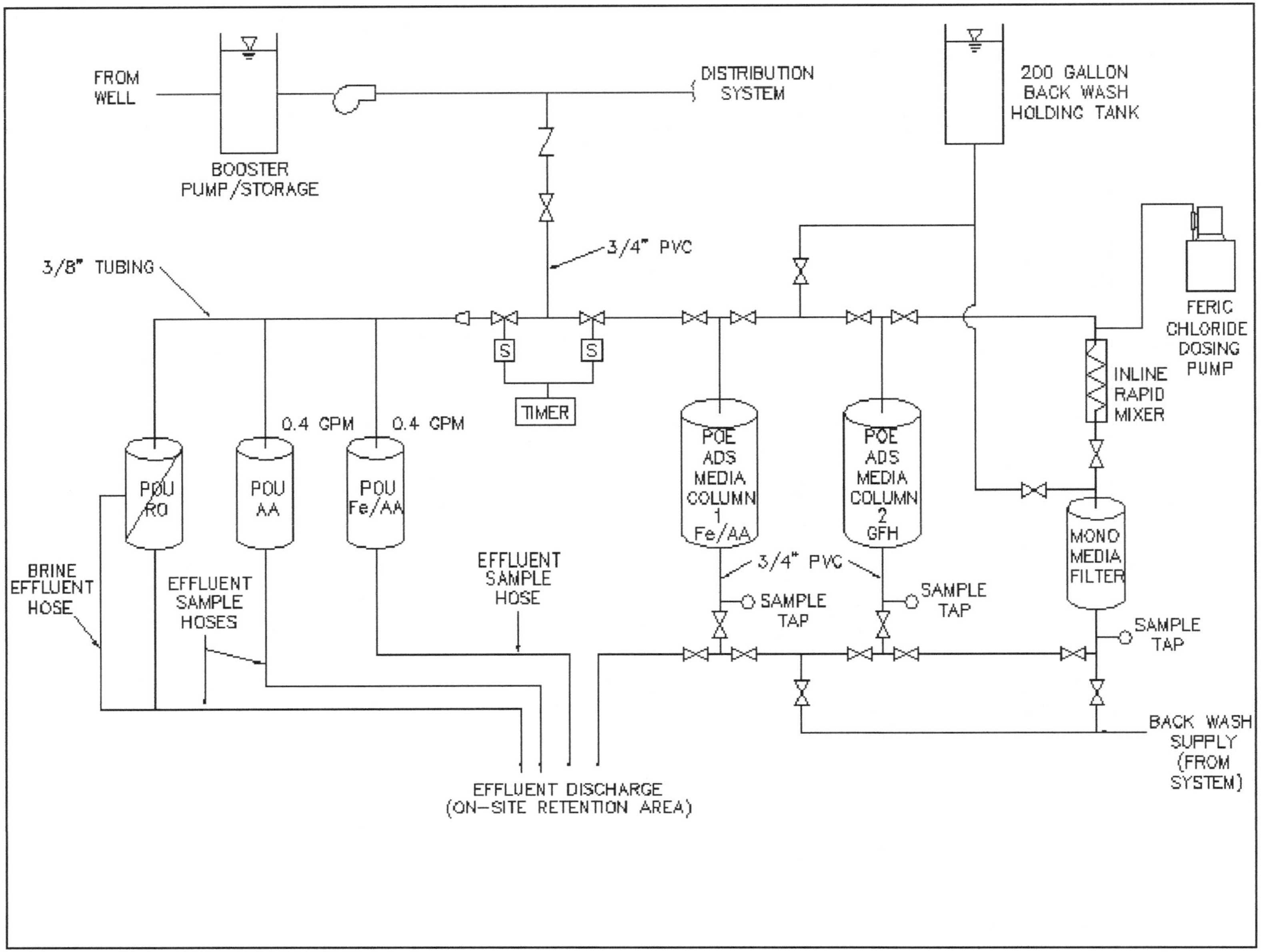

Figure D.1 Process schematic for Sun City West, Arizona

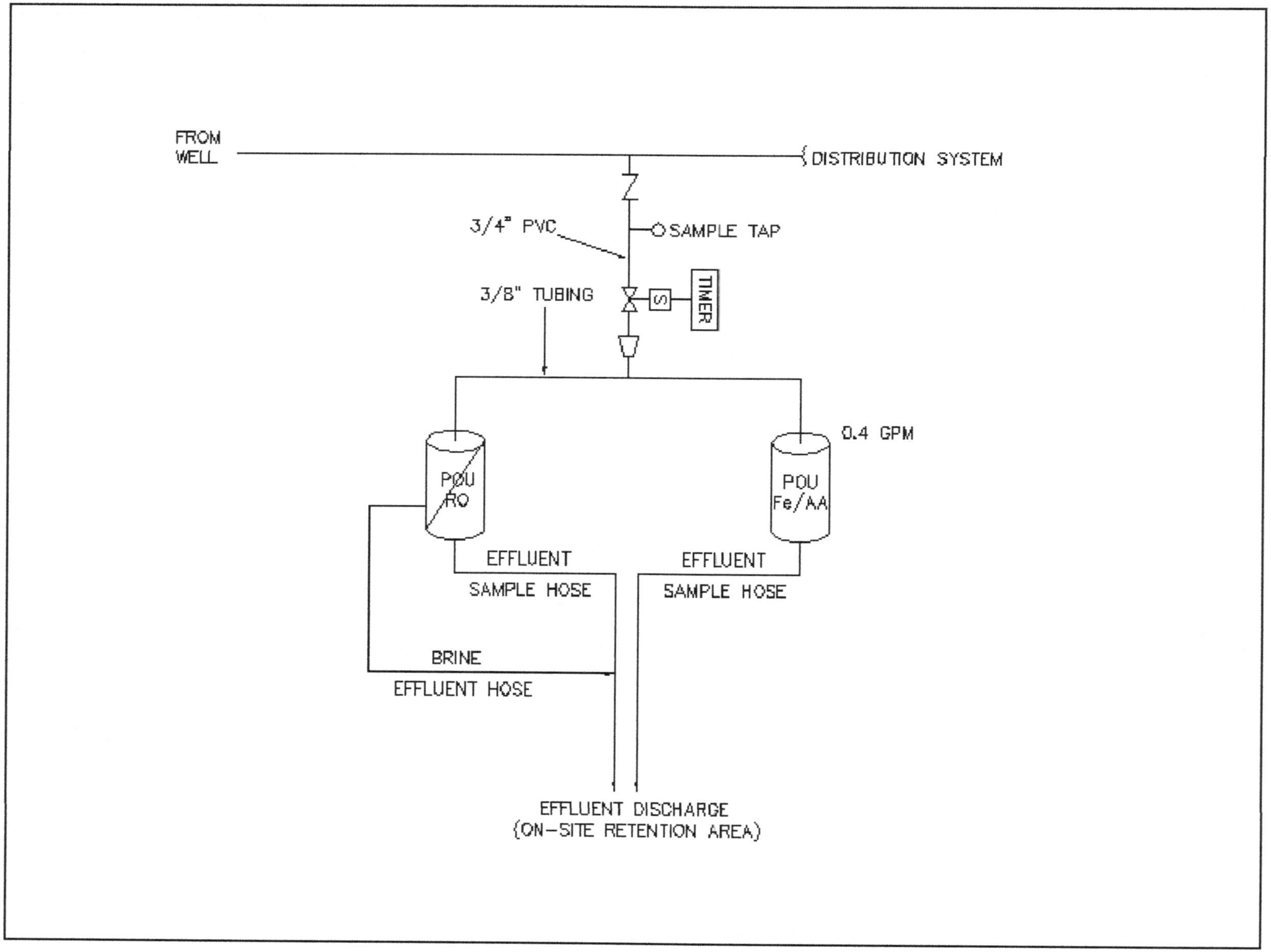

Figure D.2 Process schematic for Unity, Maine

Metro Water, Well 3, Tucson, Ariz.

Most of the arsenic at Well 3 of Metro Water in Tucson, Arizona is of As^{+5} form. The water from Well 3 has moderate concentrations of sulfate and silica. POU RO and POU AA will be tested at this location. Additionally, two 3 to 5-gpm 12-inch POE GFH and Fe-AA media units will also be tested at this site (both in intermittent and continuous mode operation). This will provide documentation of performance, costs and operator expertise for POE systems. The schematic for this site is shown in Figure D.3.

Stagecoach, Nev.

Two 3 to 5-gpm 12-inch POE columns using GFH and Fe-AA media will be tested at this site (both in intermittent and continuous mode operation). The schematic for this site is shown in Figure D.4.

Carson City, Nev.

A POU Mn-AA system and POU GFH system were installed at this site. The well is operated seasonally and was is shut down for several months. The POU cartridges will be operated based on the operation of the well.

4.0 TESTING PROTOCOL

4.1 Phase A

4.1.1 POU Units Testing

Phase A of the screening test will be designed according to the NSF protocol for POU devices (Standard 53 and Standard 58). The units will be operated using 40 minutes on/ 40 minutes off operating cycles, 16 hours per day, followed by an 8 hour rest under pressure. This will be performed for one to two weeks and it is expected that most of the adsorption media will be exhausted at the end of the testing period.

The Aquabind MP (Mn-AA) media with the 4.5" × 20" cartridge is estimated by the manufacturer to treat 1,500 gallons (1,100 BVs) before 10 ppb breakthrough. With a 4.5" × 10" cartridge, its estimated capacity is 1000 gallons (1450 BVs) prior to breakthrough. In previous tests conducted in the Phoenix, AZ area by NCS with a 2.5 minute EBCT, AA media has achieved 8,400 BVs prior to 10 ppb arsenic breakthrough (raw water arsenic 17 ppb). This would amount to approximately 5,780 gallons treated for the 4.5" × 10" cartridge and 11,600 gallons treated for the 4.5" × 20" cartridge, prior to 10 ppb breakthrough (raw water arsenic 17 ppb). The GFH media has achieved 34,000 to 85,000 BVs in pilot tests in Germany. Previous studies by NCS in Arizona, GFH with a 2.5 minute EBCT produced 42,100 BVs to 10 ppb arsenic breakthrough. The number of gallons treated with GFH would be approximately 29,000 gallons and 58,100 gallons for the 4.5" × 10" and 4.5" × 20" cartridges respectively, for waters with pH levels below 8.

This initial testing, summarized in Tables D.2 and D.3, will assist the team in comparing the results obtained from the three month screening to results obtained from manufacturers on challenge waters. For POU units, morning (9:00 am) and afternoon (12:00 noon) samples will be

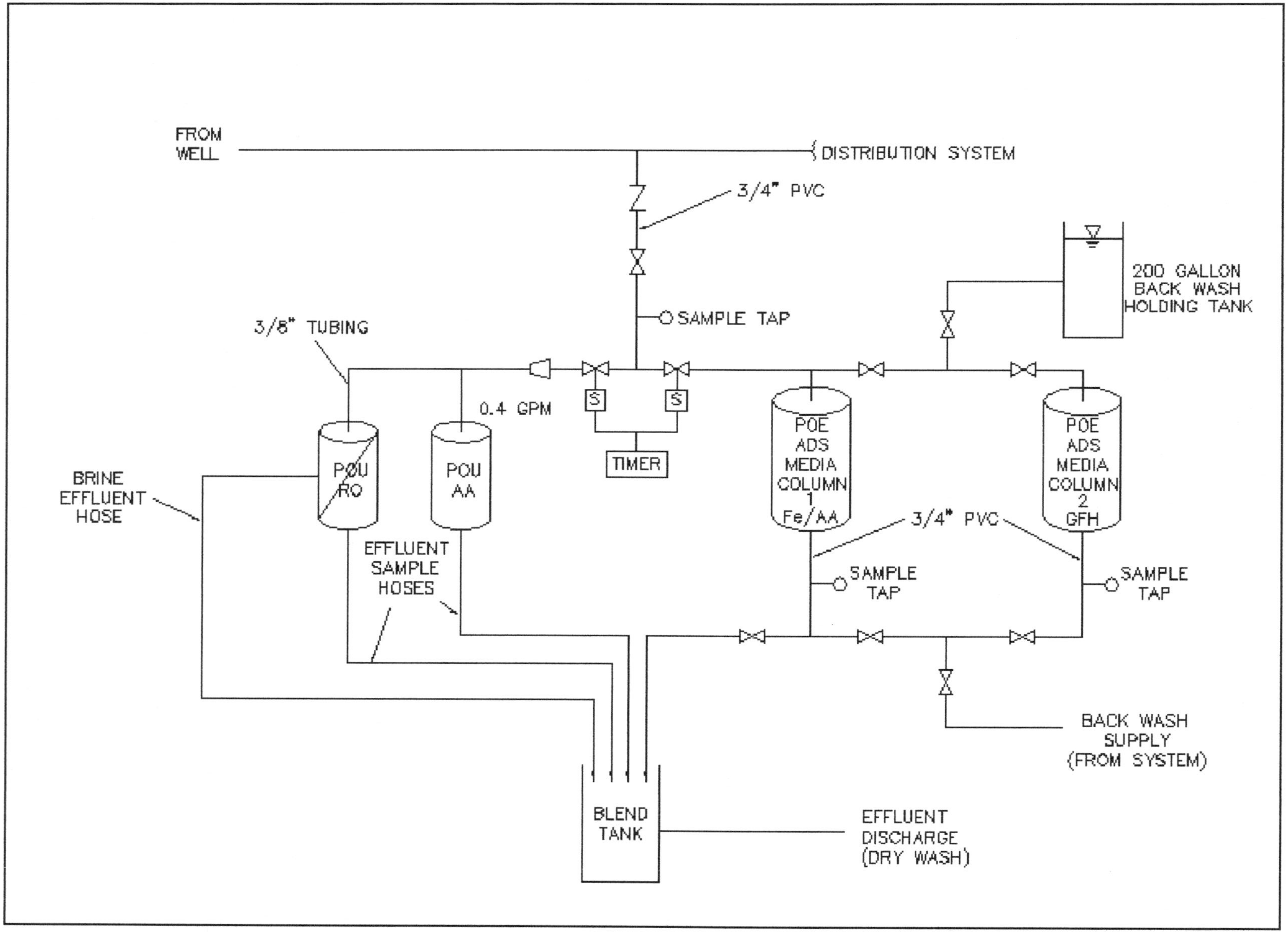

Figure D.3 Process schematic for Metro Water Tuscon, Ariz.

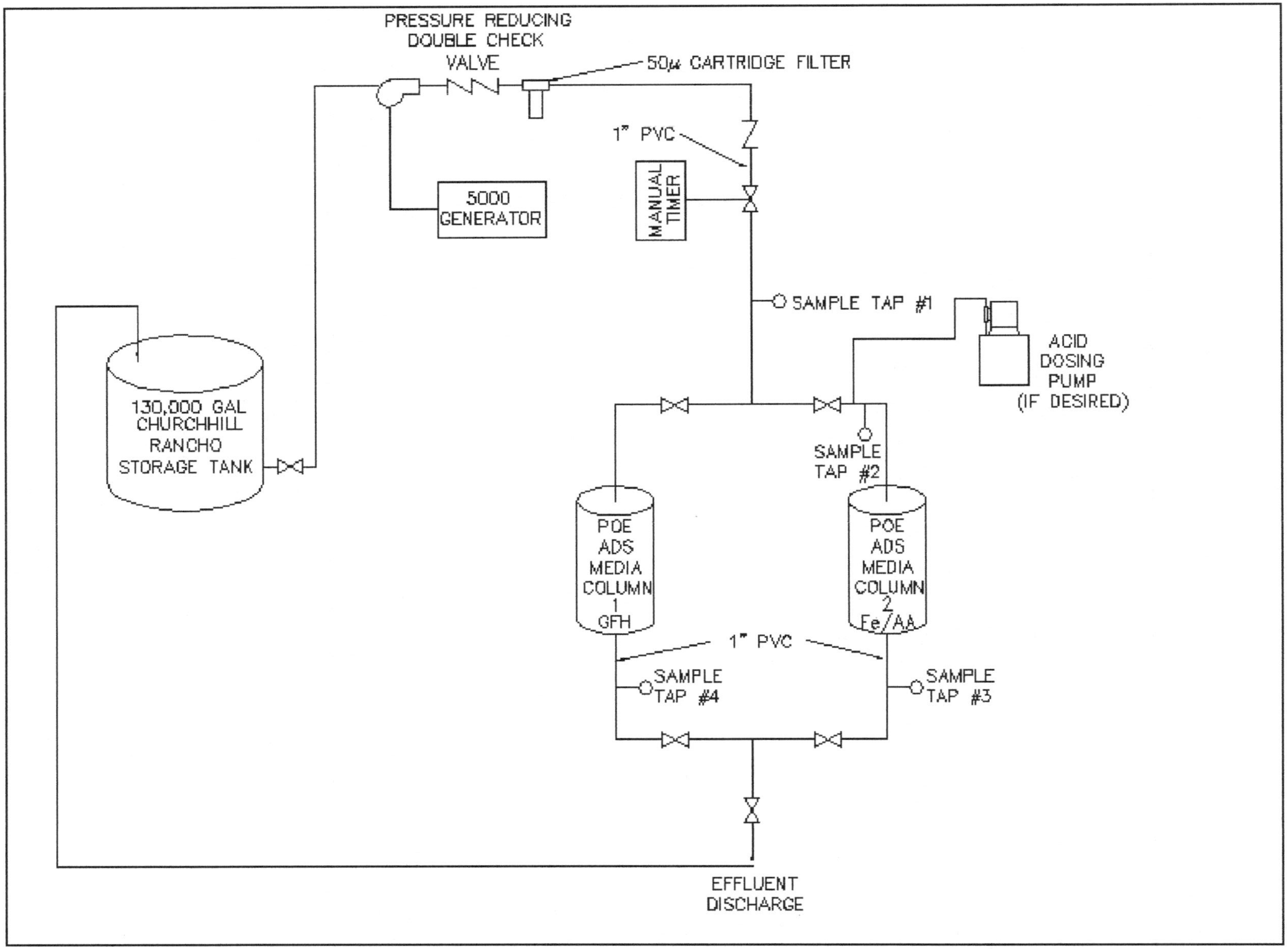

Figure D.4 Process schematic for Stagecoach, Nev.

Table D.2
Initial testing of POU units during Phase A

Period	Operating cycle	Duration per day	Sample times (3 per week)
First 1-2 weeks	40 minutes on/ 40 minutes off	16 hours followed by 8 hour rest under pressure	9:00 AM 12:00 PM

Table D.3
Sample times and parameters sampled during Phase A

Sample times	Raw	Treated effluent
9:00 AM	Arsenic, TDS, alkalinity, hardness, conductivity, sulfate, nitrate, iron, manganese, fluoride, silica, phosphate, HPC and Total coliform (weekly) Temperature, pH, chlorine residual (3 per week)	Arsenic, pH, chlorine residual, conductivity, and silica (3 days per week)
12:00 PM	—	Arsenic, pH, chlorine residual, conductivity, and silica

taken. One raw water sample per week will be analyzed for arsenic, TDS, alkalinity, hardness, conductivity, sulfate, nitrate, iron, manganese, fluoride, phosphate and silica. Raw water temperature, pH, and disinfectant residual will be analyzed three times per week. The treated water will be analyzed for arsenic, pH, chlorine residual, conductivity, and silica three times per week.

4.1.2 Testing of POE Units

The 12-inch POE columns will be set to run on a different operational schedule than the POU units. They will be operated for 16 hours on and 8 hours daily throughout for the 12-week testing period, and possibly longer, if necessary, to achieve breakthrough. During the first two weeks of testing, the POE units will be operated without pH adjustment. At this stage, pH adjustment may be made, as necessary, to optimize arsenic removal efficiency. The GFH columns will be operated at a 2.5-min EBCT and the Fe/AA columns will be operated at a 5-min EBCT.

The sampling required for the POE columns is listed in Table D.4. It is noted that only one set of the raw water samples will be taken at the site per week, for both POU and POE samples. The raw water samples have been included in Tables D.4 and D.5 for the POU units and the POE units because some of the sites only have POU units and some sites only have POE units.

Table D.4
POE unit samples and analyses*

Technology	Operational schedule	Raw	Treated effluent
POE Columns: Fe-AA and GFH	Phase A: 16 hours on, 8 hours off Phase B: continuous	Parameters and frequency same as raw samples for POU units (one set per test site) See Table D3.	Arsenic, pH, chlorine residual, conductivity, and silica (3 days per week)

* Stagecoach, NV; Metro Water, AZ; and Sun City West, AZ

Table D.5
POU testing schedule during Phase B

Period	Operating times	Duration (min)	Sampling (weekly)
1st period (2 weeks), Weeks 1-2	6:00 am	2	200 mL first flush (total coliform and HPC) sample
	8:30 am	5	
	11:30 am	5	1 L (arsenic, TDS, alkalinity, hardness, conductivity, sulfate, nitrate, iron, manganese, fluoride, silica, phosphate temperature, pH, disinfectant residual)
	5:30 pm	2	
	6:30 pm	5	
	9:30 pm	3	
2nd period (1 week), Week 3	No flow		1 L before shut down (arsenic, TDS, alkalinity, hardness, conductivity, sulfate, nitrate, iron, manganese, fluoride, silica, phosphate, temperature, pH, disinfectant residual),
3rd period (7 weeks), Weeks 4-10	6:00 am	2	200 mL first flush (total coliform and HPC)
	8:30 am	5	
	11:30 am	5	1 L (arsenic, TDS, alkalinity, hardness, conductivity, sulfate, nitrate, iron, manganese, fluoride, silica, phosphate, temperature, pH, disinfectant residual)
	5:30 pm	2	
	6:30 pm	5	

Table D.6
Laboratory analytical methods for arsenic

Analyte	Container	Sample type	Preservation	Maximum storage recommended	Preferred method(s)
Arsenic	Plastic, Glass	Grab	HNO_3 to pH <2, refrigerate and store at 4°C	28d	SM 3113 (Electrothermal Atomic Absorption Spectrometric method) SM 3114 B (Manual Hydride generation/Atomic absorption Spectrometric method) SM 3500-As B (Silver Diethyldithiocarbamate Method) SM 3120 B (ICP/AES) SM 3125 B (ICP/MS)

4.2 Phase B

Phase B of the testing procedure consists of a 10-week screening test for the POU units. The testing protocol for the POU units will be designed to mimic actual household operation of the POU units. Long periods of quiescence will also be used to simulate vacation periods. The daily operational schedule for the POU units is presented in Table D.5. The operating start times and duration will be programmed into the timers at each site.

In addition to the parameters included in Phase A, total coliform and HPC samples will be collected weekly from the influent and effluent units. To determine the potential for leaching, the Material Safety Data Sheets (MSDS) for the adsorption media were examined and iron, aluminum and manganese were identified as possible leachates from the media from the Mn-AA and AA media. Aluminum and manganese were identified as potential leachates from the POU units. Samples from the effluent of the POU units and POE units will be taken once per month for the parameters identified for each media. However, it is emphasized that from more than a year of data collection on leachates from adsorption media, NCS found minimal potential for leaching from granular iron and alumina media.

5.0 QA/QC

Laboratory analytical methods for arsenic are shown in Table D.6 and analytical methods for the other parameters sampled from the POU units are listed in Table D.7.

Table D.7
Laboratory analytical methods for other parameters

Analyte	Container	Sample type	Preservation (Refrigerate; 4°C)	Maximum storage recommended	Preferred method(s)
Silica	Plastic	Grab	Refrigerate, do not freeze	28 days	EPA 200.7
Sulfate	Plastic, Glass	Grab	Refrigerate	28 days	EPA 300
Alkalinity	Plastic, Glass	Grab	Refrigerate	24 hours	SM 2320 B
Conductivity	Plastic, Glass	Grab	Refrigerate	48 hours	SM 2510 B
pH	Plastic, Glass	Grab	Analyze immediately	15 minutes	SM 4500-H^+ B
Hardness	Plastic, Glass	Grab	HNO_3 to pH<2 or H_2SO_4 to pH<2	6 months	SM 2340 C
Iron	Plastic, Glass	Grab	HNO_3 to pH<2 or H_2SO_4 to pH<2	6 months	SM 3120 B
Manganese	Plastic, Glass	Grab	HNO_3 to pH<2 or H_2SO_4 to pH<2	6 months	SM 3120 B
Fluoride	Plastic	Grab	None required	28 days	EPA 1010 SM 4500 F^-
Total coliform	Plastic, Glass	Grab	Cool, 4°C, 0.008% $Na_2S_2O_3$	6 hours	SM 9221 B
HPC	Plastic, Glass	Grab	Cool, 4°C, 0.008% $Na_2S_2O_3$	6 hours	SM 9215
TDS	Plastic, Glass	Grab	Refrigerate	7 days	SM 2540 C

6.0 SAMPLE TAGGING AND CHAIN OF CUSTODY PROCEDURES

A sample tagging procedure will be followed for all the analytical samples according to the following scheme:

Project Name-Utility Code-Technology Code-Sample Location

<u>Project Name</u> - A2730

Utility Codes

Sun City West, Well 28 (Phoenix Metro Area), Ariz	SCW
Sandy Stream Water System, Unity, Maine	ME
Well 3, Metro Water, Tucson, Ariz.	TU
Stagecoach, Nev.	ST
Carson City, Nev.	CC

Technology Codes (contactors will be clearly labeled for each media that is contained within)

POU-RO	RO
POU-Aquabind MP (Mn-AA)	MP
POU-Aquabind EP (AA)	EP
POE-12" column-Fe-AA	FE-AA
POE-12" column-GFH	GFH

Sample Location

Raw Water	R
Effluent	E

For example, an effluent water sample collected from the RO unit at Sun City West, Ariz. would be labeled "A2730-SCW-RO-E."

Each utility will perform their own sampling of test units with guidance and oversight from the research team. Samples from Metro Water, Tucson, Ariz.; Unity, Maine and Sun City West, Ariz. sites will be delivered to a certified private laboratory for analyses. Samples from the Stagecoach, Nev. site will be sent to the Nevada State laboratory for analysis.

APPENDIX E
POST-INSTALLATION SURVEY

Table E.1
POU installation form

Customer name:

Customer address:

Customer phone #:

Number of people in Home:

Type of Unit Installed: Reverse Osmosis ☐
EP Adsorption Cartridge ☐
MP Adsorption Cartridge ☐
SP-33 Adsorption Cartridge ☐
Was a flow totalizer installed ☐

	Highly Unfavorable	Somewhat Unfavorable	Don't Care	Somewhat Favorable	Favorable	Highly Favorable
Customer's attitude about the overall program	0	1	2	3	4	5
Customer's attitude about the installation activities	0	1	2	3	4	5
Customer's attitude about the sampling program	0	1	2	3	4	5
How important was it to the customer that they drink arsenic free water?	0	1	2	3	4	5
Where they concerned with how long the installation would take?	0	1	2	3	4	5
Did they have different attitudes toward contractors in comparison to utility personnel?	0	1	2	3	4	5

Was the under sink area already cleaned out?	no	yes
Were they helpful in cleaning it out?	no	yes
Were they concerned with loss of storage space?	no	yes
Were they concerned with the extra hole in the sink?	no	yes
Were they concerned with removal after testing?	no	yes
Were they concerned with leaks and whom to call?	no	yes
Were they concerned about exposure through skin contact?	no	yes
Did they watch you install the POU unit?	no	yes
Did they ask questions about the POU unit?	no	yes
Was the RO line run to the icemaker/refrigerator?	no	yes
Were they interested in removing T&O , TDS, or Chlorine?	no	yes
Did they inquire about cost?	no	yes
Were there access problems (availability, animals, lack of response)?	no	yes

Remarks:

Reaction of people with existing POU units (impacts on performance) willingness to participate

Consumption rates 1-5 glasses per person per day?

Cooking use 1-10 glasses?

Use of other types for drinking?

APPENDIX F
EXIT SURVEY

Table F.1
In-home testing exit survey

Customer name: ______
Customer address: ______
Customer phone #: ______
Number of people in home: ______

Type of unit installed: Reverse osmosis ☐
MP adsorption cartridge ☐
Was a flow totalizer installed? ☐ Volume?

Customer's attitude about:
-- Overall program? Ranking - ______
-- Sampling program? Ranking - ______
-- Comments:

Did the treated water taste better than tap water? No Yes
-- Comments: (Ranking: __________)

Did customer notice any other benefits of using the POU? No Yes
-- Comments:

Ranking Key:
0 - Highly Unfavorable
1 - Somewhat Unfavorable
2 - Don't Care
3 - Somewhat Favorable
4 - Favorable
5 - Highly Favorable

Is it more important that the water taste good or be arsenic free? Taste 1 2 3 4 5 Arsenic Free

Did you use other taps for drinking? No Yes
-- Concerned about being able to drink from only one tap? No Yes
Estimated Daily drinking water consumption rates:
-- Drinking water volume? ______
-- Cooking water volume? ______
-- Volume of water consumed from POU? ______
-- Comments:

Who would the customer perfer to conduct the sampling/maintenance? Utility Employee Private Contractor No Difference
-- Comments:

Were there any problems with plumbing or leaks during the test period? No Yes
-- If yes, comments:

Does the customer plan to keep the POU unit? No Yes
-- If no, concerned with removal of unit after testing? No Yes

Remarks / other concerns?

REFERENCES

Bell, F.A. 1988. POU/POE Point of View. In *Proc. of the Conference on Point-of-Use Treatment of Drinking Water*. Cincinnati, Ohio: USEPA.

Bell, F.A., et al. 1984. Studies on Home Water Treatment Systems. *Jour. AWWA*, 76(4):126.

Bellack, E. 1971. Arsenic Removal from Potable Water. *Jour. AWWA*, 63(7):454–458.

Brandhuber, P., and G. Amy. 1998. Alternative Methods for Membrane Filtration of Arsenic from Drinking Water. *Desalination*, 117:1–10.

Cadmus Group, Inc. 1998. Cost Evaluation of Small System Compliance Options—Point-of-Use and Point-of-Entry Treatment Units. Prepared for USEPA Office of Groundwater and Drinking Water.

California Department of Health Services (CADHS). 2000. Requirements for Point of Entry (POE) Systems on Constructed Conveyance Water Systems (Irrigation Districts). Drinking Water Program; Certified Water Treatment Device Directory. Sacramento, Calif.

Clifford, D., and C.C. Lin. 1991. Arsenic (III) and Arsenic (V) Removal from Drinking Water in San Ysidro, New Mexico. EPA/600/2-91:11.

DeFilippi, J.A., and J.H. Baier. 1987. Point-of-Use and Point-of-Entry Treatment on Long Island. *Jour. AWWA*, 79(10):76.

Den, J. 1999. Small Water Systems Need You. [Online]. Available: <http://www.watertechonline.com/article/asp?IndexID=5220801>.

Fox, K.R. 1989. Field Experience with Point-of-Use Treatment Systems for Arsenic Removal. *Jour. AWWA*, 2:94–101.

Fox, K.R., and T.J. Sorg. 1987. Controlling Arsenic, Fluoride and Uranium by Point-of-Use Treatment. *Jour. AWWA*, 79(10):81–84.

Frank, P., and D. Clifford. 1986. Arsenic (III) Oxidation and Removal from Drinking Water. EPA-600-52/86-021.

Geldreich, E.E., R.H. Taylor, J.C. Blannon, and D.J. Reasoner. 1985. Bacterial Colonization of Point-of-Use Treatment Devices. *Jour. AWWA*, 77(2):72.

Gilles, G., and S. Odom. 2000. Arsenic Reduction Challenges in India and Bangladesh. *Water Conditioning and Purification*, Volume 45, No. 6.

Gupta, S.K., and K.Y. Chen. 1978. Arsenic Removal by Adsorption. *Jour. WPCF*, 3(3):493.

Hamill, B. 1987. POU/POE—Point of View—Association of State Drinking Water Administrators (ASDWA). Bureau of Safe Drinking Water. New Jersey Department of Environmental Protection.

Hecht, P.M., D.J. Hiltebrand, and J.D. Lowry. 1992. Arsenic Removal by Anionic Exchange. In *Proc. of the AWWA Annual Conference*. Denver, Colo.: AWWA.

Idaho Department of Environmental Quality (IDEQ). 2000. Guidelines for Approving Point-of-Use Treatment in Idaho Public Water Systems. Guidance document. Boise, Idaho.

Knowles, F.C., and A.A. Benson. 1983. *Trends Biochem. Sci*, 8:178–180.

Korte, N.E., and Q. Fernando. 1991. *Crit. Rev. Environ. Control*, 21:1–39.

Narasimhan Consulting Services, Inc. (NCS, Inc.) 2000. Arizona Arsenic Master Plan, Part 3—Compliance Options. Prepared for Arizona Department of Environmental Quality.

NCS, Inc. 2001a. Arsenic Study for Surface Water and Ground Water. City of Phoenix Project No. WS85700023-S.

NCS, Inc. 2001b. Membrane Testing on Phoenix Area Surface Water. Part of Lake Pleasant Water Treatment Plant: Water Quality Testing Study, City of Phoenix Project No. WS85350002.

Pagliano, T. 1999. NSF Revision of RO Standard Could be Finalized Soon. *Water Technology*, 8:52.

Pennsylvania Department of Environmental Protection (PADEP). 2003. Safe Drinking Water POE/POU Regulations. Bureau of Water Supply and Wastewater Management; Chapter 9. [Online]. Available: <http://www.pacode.com/secure/data/025/chapter109/chap109toc.html>

Raucher, R.S., M.L. Hagensted, J. Cotruvo, R. Narasimhan, and R. Regunathan. 2004. *Conventional and Unconventional Approaches to Water Service Provision*. Denver, Colo.: AwwaRF and AWWA.

Reasoner, D.J., J.C. Blannon, and E.E. Geldrich. 1987. Microbial Characteristics of Third Faucet Point-of-Use Devices. *Jour. AWWA*, 79(10):60–66.

Rogers, K. 1990. Community Demonstration of POU Systems Removal of Arsenic and Fluoride: San Ysidro, New Mexico. In *Proc. of the Conference on Point-of-Use Treatment of Drinking Water*. Cincinnati, Ohio: USEPA.

Taylor, R.H., M.J. Allen, and E.E. Geldreich. 1979. Testing Home Use Carbon Filters. *Jour. AWWA*, 71(1):577.

Thomson, B.M., T.J. Cotter, and J.D. Chwirka. 2002. Technical Note: Design and Operation of a Point-of-Use Treatment System for Arsenic Removal. *Jour. Environ. Engr.*

Thomson, B.M., K. Fox, and M. O'Grady. 2000. Arsenic Removal by Point-of-Use Treatment Systems at San Ysidro, New Mexico. In *Proc. of the AWWA Annual Conference*. Denver, Colo.: AWWA.

Thomson, B.M., and M. O'Grady. 1995. Evaluation of Point-of-Use Water Treatment Systems, San Ysidro, New Mexico. Cincinnati, Ohio: EPA-600/R3-95-113.

Tobin, R.S. 1987. Testing and Evaluation of Point-of-Use Treatment Devices in Canada. *Jour. AWWA*, 10:42.

U.S. Environmental Protection Agency (USEPA). 2001. 40 CFR Parts 9, 141 and 142. National Primary Drinking Water Regulations; Arsenic and Clarifications to Compliance and New Source Contaminants Monitoring; Final Rule. *Fed. Reg.*, 66(14): 6976–7066.

USEPA. 2002. Guidance for Implementing a Point-of-Use or Point-of-Entry Treatment Strategy for Compliance with the Safe Drinking Water Act; Revised Final Draft. [Online]. Available: <http://www.epa.gov/safewater/standard/draftguid_pou_poe.pdf>

USEPA. 2004. Centrally Managed POU Treatment for Compliance with the Arsenic Rule. [Online]. Available: <http://www.epa.gov/safewater/dwa/pdfs/webcast/presentations/webcast 4/bellen_pou_case_study.pdf>

World Health Organization (WHO). 2002. *The World Health Report 2002—Reducing Risks, Promoting Healthy Life*. Geneva: WHO. [Online]. Available: http://www.who.int/whr/2000/en/> Accessed October 2003.

ABBREVIATIONS

AA	activated alumina
ANSI	American National Standards Institute
ASDWA	Association of State Drinking Water Administrators
AWWA	American Water Works Association
AwwaRF	Awwa Research Foundation
BVs	bed volumes
CADHS	California Department of Health Services
CFU/mL	colony forming units per milliliter
DBPs	disinfection by-products
DWTU	Drinking Water Treatment Unit
EBCT	empty bed contact time
EDR	electrodialysis reversal
EPA	Environmental Protection Agency
Fe-AA	iron enhanced activated alumina
ft^3	cubic feet
GAC	granular activated carbon
GFH	granular ferric hydroxide
GIM	granular iron media
gpd	gallons per day
gpm	gallons per minute
HPC	heterotrophic plate count
IDEQ	Idaho Department of Environmental Quality
IX	ion exchange
LAAFP	Los Angeles Aqueduct Filtration Plant
LADWP	Los Angeles Department of Water and Power
MCL	maximum contaminant level
mg/L	milligrams per liter
MGD	million gallons per day
Mn-AA	manganese coated activated alumina
μg/L	micrograms per liter

NCS	Narasimhan Consulting Services, Inc.
NOM	natural organic matter
NSF	National Sanitation Foundation International
PADEP	Pennsylvania Department of Environmental Protection
POE	point of entry
POU	point of use
ppb	parts per billion
psi	pounds per square inch
RO	reverse osmosis
SDWA	Safe Drinking Water Act
TCLP	toxic characteristic leaching procedure
TDS	total dissolved solids
TFC	thin film composite
UL	Underwriters Laboratories Inc.
UV	ultraviolet
VOCs	volatile organic compounds
WHO	World Health Organization
WQA	Water Quality Association

IWA Publishing's authorised EU representative for General Product Safety Regulations is Diane D'Arras, 15 rue Duret, 75116 Paris, France, e-mail: safety@iwap.co.uk.

Printed and bound by CPI Group (UK) Ltd, Croydon, CR0 4YY
17/09/2026
02218799-0003